WJEC
Mathematics
for A2 Level
Pure & Applied
Practice Tests

Stephen Doyle

Illuminate Publishing

Published in 2019 by Illuminate Publishing Ltd, P.O Box 1160, Cheltenham, Gloucestershire GL50 9RW

Orders: Please visit www.illuminatepublishing.com
or email sales@illuminatepublishing.com

British Library Cataloguing in Publication Data

A catalogue record for this book is available from the British Library

ISBN 978-1-911208-56-3

Printed by Ashford Colour Press, Gosport

07.19

The publisher's policy is to use papers that are natural, renewable and recyclable products made from wood grown in sustainable forests. The logging and manufacturing processes are expected to conform to the environmental regulations of the country of origin.

Editor: Geoff Tuttle
Cover design: Neil Sutton
Text design and layout: GreenGate Publishing Services, Tonbridge, Kent

Photo credits

Cover: Klavdiya Krinichnaya/Shutterstock; **p5** Gajus/Shutterstock; **p8** Maren Winter/Shutterstock; **p17** gorosan/Shutterstock; **p21** Yury Zap/ Shutterstock; **p26** Chad Bontrager/Shutterstock; **p33** Jakkrit Orrasri/Shutterstock; **p36** Kirill Neiezhmakov/Shutterstock; **p45** Mvolodymyr/Shutterstock; **p50** Dark Moon Pictures/Shutterstock; **p55** Russell Parry; **p61** iMoved Studio/ Shutterstock; **p68** Elana Erasmus/Shutterstock; **p71** Serhii Moiseiev /Shutterstock; **p74** drpnncpptak/Shutterstock; **p78** Alexey Broslavets/Shutterstock; **p84** Aspen Photo/Shutterstock; **p88** Martin Lisner/Shutterstock; **p91** and **p97** panitanphoto/ Shutterstock.

Acknowledgements

The author and publisher wish to thank Sam Hartburn for her help and careful attention when reviewing this book.

Contents

Questions

Answers

Introduction

This Practice Tests book is designed to be used alongside the WJEC Mathematics for A2 Level Pure and Applied textbooks. The book follows the exact topic order in the book, which is the same as the topic order in the WJEC specification.

The main purpose of this book is to build confidence in the topics by providing carefully graded questions, many of which are similar to the sorts of questions you might get in the actual examination.

Here are some of the features of the Practice Tests Book:

- Facts and formulae at the start of each topic making it easy for you to see what you should already know.
- Topic by topic carefully graded questions containing space for you to write your answers.
- Full answers at the back to all the questions with full explanations and explanations of alternative ways of solving the same problem.
- Tips on answering the questions to maximise your marks.
- Unstructured questions are provided which are a new feature of the specification.
- Specimen test papers for you to try.

1 Proof

Essential facts and formulae

Facts

There are several methods for proving or disproving a mathematical statement.

In many examination questions, you will be required to use a named proof and in A2 Pure you will need to use a new proof called proof by contradiction.

Proof by contradiction – where you assume that a conjecture is false and then show that this assumption leads to a contradiction.

Questions

1 Use proof by contradiction to prove the following proposition:

If a is a real integer and a^2 is even, then a is even. [4]

2 Use proof by contradiction to prove that there are no positive integer solutions to the equation $x^2 - y^2 = 1$. [6]

3 Prove by contradiction the following proposition

When x is real and $\frac{\pi}{2} \le x \le \pi$, $\sin x - \cos x \ge 1$. [6]

4 Complete the following proof by contradiction to show that:

$\sin \theta + \cos \theta \le \sqrt{2}$ for all values of θ.

The first two lines of the proof are given below:

Assume that there is a value of θ for which $\sin \theta - \cos \theta > \sqrt{2}$

Then squaring both sides, we have ... [5]

5 Prove by contradiction the following proposition:

When x is real and $x \neq 0$,

$$\left| x + \frac{1}{x} \right| \geq 2$$

The first two lines of the proof are given below:

Assume that there is a real value of x such that $\left| x + \frac{1}{x} \right| < 2$

Then squaring both sides, we have: [3]

6 Prove by contradiction the following proposition:

When x is real and positive,

$$4x + \frac{9}{x} \geq 12$$

The first line of the proof is given below:

Assume that there is a positive and real value of x such that

$$4x + \frac{9}{x} < 12 .$$ [4]

2 Algebra and functions

Essential facts and formulae

Facts

Partial fractions

Here are two of the main cases:

Where there are no repeating linear factors $\left(\text{i.e. } \dfrac{4x-5}{(3x+1)(x-1)} = \dfrac{A}{(3x+1)} + \dfrac{B}{(x-1)}\right)$

Where there are repeating linear factors $\left(\text{i.e. } \dfrac{3x^2+2x-1}{(x+4)(x+2)^2} = \dfrac{A}{x+4} + \dfrac{B}{(x+2)^2} + \dfrac{C}{x+2}\right)$

Functions

Function – relation between a set of inputs and outputs so that each input is related to just one output.

Domain – set of input values

Range – set of output values

One-to-one function – one output value corresponds to a single input value

Composite functions – applying two functions in succession. $fg(x)$ means performing g and then f.

Finding an inverse function (i.e. f^{-1}) – check first that the function is a one-to-one function otherwise the inverse does not exist. Let y equal the function and rearrange so x is the subject of the equation. Replace x on the left with $f^{-1}(x)$ and on the right replace all occurrences of y with x.

Domain and range of inverse functions – the range of $f^{-1}(x)$ is the same as the domain of $f(x)$ and the domain of $f^{-1}(x)$ is the same as the range of $f(x)$.

The modulus function – the modulus of x is written as $|x|$ and means the numerical value of x (ignoring the sign). So whether x is positive or negative, $|x|$ is always positive (or zero).

Graphs of modulus functions – first plot the graph of $y = f(x)$ and reflect any part of the graph below the x-axis in the x-axis. The resulting graph will be $y = |f(x)|$.

Combinations of transformations – if a graph of $y = f(x)$ is drawn, then the graph of $y = f(x - a) + b$ can be obtained by applying the translation $\begin{pmatrix} a \\ b \end{pmatrix}$ to the original graph.

If a graph of $y = f(x)$ is drawn, then the graph of $y = af(x - b)$ can be obtained by applying the following two transformations in either order: a stretch parallel to the y-axis with scale factor a and a translation of $\binom{b}{0}$.

If a graph of $y = f(x)$ is drawn, then the graph of $y = f(ax)$ can be obtained by scaling the x-values by $\frac{1}{a}$.

The function e^x and its graph

$y = e^x$ is a one-to-one function and so has an inverse (i.e. $y = \ln x$) and intersects the y-axis at $y = 1$, and it has the x-axis as an asymptote

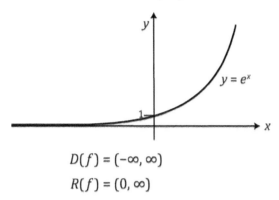

$$D(f) = (-\infty, \infty)$$
$$R(f) = (0, \infty)$$

The function $\ln x$ and its graph

$y = \ln x$ is a one-to-one function and so has an inverse (i.e. $y = e^x$). It cuts the x-axis at $x = 1$ and has the y-axis as an asymptote

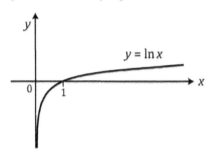

$$D(f) = (0, \infty)$$
$$R(f) = (-\infty, \infty)$$

The functions $y = \ln x$ and $y = e^x$ are inverse functions.

So, the graphs of $y = \ln x$ and $y = e^x$ are reflections of each other in the line $y = x$.

Also, $e^{\ln x} = x$ and $\ln e^x = x$.

Formulae

$$\frac{ax + b}{(cx + d)(ex + f)} \equiv \frac{A}{cx + d} + \frac{B}{ex + f} \qquad \text{(i)}$$

$$\frac{ax^2 + bx + g}{(cx + d)(ex + f)^2} \equiv \frac{A}{cx + d} + \frac{B}{ex + f} + \frac{C}{(ex + f)^2} \qquad \text{(ii)}$$

In both cases, clear the fractions and choose appropriate values of x. In (ii), an equation involving coefficients of x^2 may be used.

Questions

 Simplify each of the following expressions:

(a) $\dfrac{x^2 - 4}{x^2 + x - 2}$ [1]

(b) $\dfrac{x^2 + 2x + 1}{3x^2 + 12x + 9}$ [2]

2 If the algebraic fraction $\dfrac{5x^2 + 6x + 7}{(x - 1)(x + 2)^2} \equiv \dfrac{A}{(x + 2)^2} + \dfrac{B}{x + 2} + \dfrac{C}{x - 1}.$

Find the integers A, B and C. [4]

3 Given that $f(x) = \dfrac{3x}{(1+x)^2(2+x)}$

express $f(x)$ in terms of partial fractions. [4]

4 The diagram shows a sketch of the graph of $y = f(x)$. The graph passes through the points $(-1, 0)$ and $(7, 0)$ and has a minimum point at $(3, -6)$.

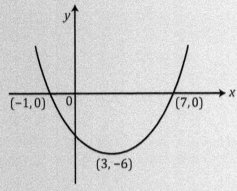

Sketch the graph of $y = -\frac{2}{3}f(x + 4)$, indicating the coordinates of the stationary point and the coordinates of the points of intersection of the graph with the x-axis. [3]

5 The functions f and g have domains $(0, \infty)$ and $\left(0, \frac{\pi}{4}\right)$ respectively and are defined by:

$$f(x) = \ln x,$$
$$g(x) = \tan x.$$

(a) (i) Write down the domain of fg.

    (ii) Write down the range of fg.   [3]

(b) (i) Solve the equation $fg(x) = -0.4$.
       Give your answer correct to two decimal places.

    (ii) Giving a reason, write down a value for k so that $fg(x) = k$ has no solution.   [3]

6 In a college with 6000 students, one student returns from the Christmas holiday with a contagious flu virus. The spread of the virus is modelled by the following function:

$$f(t) = \frac{6000}{1 + 5999e^{-0.8t}}$$

where $f(t)$ is the total number of students infected after t days.

(a) Explain why the domain of this function is $t \geq 0$.   [1]

(b) Explain the significance of $f(t)$ when $t = 0$.   [1]

(c) Find the number of students infected with the virus after 5 days.   [3]

(d) The college will cancel classes when 40% or more of the students are infected with the virus. Find after how many days the college will cancel classes, giving your answer to one decimal point.   [3]

7 Solve $|2x + 3| = x$ [3]

8 The function f has domain $(-\infty, 3)$ and is defined by:
$$f(x) = x^2 - 6x + 8$$

(a) Sketch the graph of $y = f(x)$ marking on your graph the points of intersection with each axis and the coordinates of the minimum point. [5]

(b) Write down the range of f. [2]

(c) Find an expression for $f^{-1}(x)$ and write down the domain of f^{-1}. [3]

9 The function f is defined by $f(x) = |\cos x|$ for $0 \le x \le 2\pi$

   (a) Sketch the graph of $y = f(x)$.   [2]

   (b) On a separate set of axes, sketch the graph of $y = f(x) + 1$.   [2]

10 (a) Solve $|5x - 8| = 2$. [3]

(b) (i) Sketch the graph of $y = |x|$.

(ii) The diagram below shows a sketch of the graph $y = -x^2 + 7x - 10$

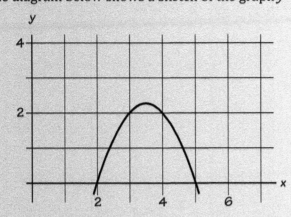

Draw a sketch of the graph $y = |-x^2 + 7x - 10|$ indicating the coordinates of all the points where the curve cuts the axes. [5]

11 The function f has domain $[1, \infty)$ and is defined by

$$f(x) = \ln(3x - 2) + 5$$

(a) Find an expression for $f^{-1}(x)$. [4]

(b) State the domain of f^{-1}. [1]

12 The functions f and g have domains $[0, \infty)$ and $(-\infty, \infty)$ respectively and are defined by

$$f(x) = e^x$$
$$g(x) = 4x^3 + 7$$

(a) Find and simplify an expression for $gf(x)$. [2]

(b) Find the domain and range of gf. [2]

(c) (i) Solve the equation $gf(x) = 18$.
Give your answer correct to three decimal places.

(ii) Giving a reason, write down a value for k so that $gf(x) = k$
has no solutions. [3]

3   Sequences and series

Essential facts and formulae

Facts

Sequence – a list of terms

Series – the sum of a certain number of terms in a sequence

Arithmetic sequence – sequence where there is a common difference between consecutive terms

Geometric sequence – sequence where there is a common ratio between consecutive terms

Convergent sequence – a sequence with a limit of infinity, so terms get larger and larger

Divergent sequence – a sequence where the nth term approaches a steady value

Periodic sequence – a sequence that repeats itself after n terms

Formulae

The binomial expansion of $(1 + x)^n$ for positive integer n

$$(1 + x)^n = 1 + nx + \frac{n(n-1)}{2!}x^2 + \frac{n(n-1)(n-2)}{3!}x^3 + \dots \text{ for } |x| < 1$$

The binomial expansion of $(a + b)^n$ for negative or fractional n

$$(a + b)^n = a^n + \binom{n}{1}a^{n-1}b + \binom{n}{2}a^{n-2}b^2 + \dots + \binom{n}{r}a^{n-r}b^r + \dots + b^n$$

$$\binom{n}{r} = {}^nC_r = \frac{n!}{r!(n-r)!}$$

The nth term of an arithmetic sequence

n^{th} term $t_n = a + (n-1)d$ where a is the first term, d is the common difference and n is the number of terms.

The sum to n terms of an arithmetic series

$$S_n = \frac{n}{2}\Big[2a + (n-1)d\Big]$$

The *n*th term of a geometric sequence

n^{th} term $t_n = ar^{n-1}$ where *a* is the first term, *r* is the common ratio and *n* is the number of terms.

The sum to *n* terms of a geometric series

$S_n = \dfrac{a(1 - r^n)}{1 - r}$ provided $r \ne 1$

The sum to infinity of a geometric series

$S_\infty = \dfrac{a}{1 - r}$ Note that for the sum to infinity to exist $|r| < 1$.

Questions

 (a) An arithmetic series has first term *a* and common difference *d*. Prove that the sum of the first *n* terms of the series is given by

$$S_n = \frac{n}{2}[2a + (n - 1)d]$$ [3]

(b) The first term of an arithmetic series is 4 and the common difference is 2. The sum of the first *n* terms of the arithmetic series is 460. Write down an equation satisfied by *n*. Hence find the value of *n*.   [3]

(c) The fifth term of another arithmetic series is 9. The sum of the sixth term and the tenth term of this series is 42. Find the first term and the common difference of the arithmetic series.   [5]

2 The nth term of a number sequence is denoted by t_n. The $(n + 1)$th term of the sequence satisfies:

$$t_{n+1} = 2t_n + 1 \text{ for all positive integers } n.$$

Given that $t_4 = 63$,

(a) evaluate t_1, [2]

(b) without carrying out any further calculations, explain why 6 043 582 cannot be one of the terms of this number sequence. [1]

3 Find the value of $\displaystyle\sum_{x=1}^{100} (3x - 2)$. [3]

4 (a) (i) Expand $\dfrac{1}{\sqrt{1+2x}}$ in ascending powers of x up to and including

the term in x^2. [3]

(ii) State the range of values of x for which your expansion is valid. [1]

(b) Use your expansion in part (a) to find an approximate value for one root of the equation:

$$\frac{6}{\sqrt{1+2x}} = 4 + 15x - x^2$$ [3]

5 (a) Expand $(1-x)^{-\frac{1}{2}}$ in ascending power of x as far as the term in $2x$. State the range of x for which the expansion is valid. [2]

(b) By taking $x = \frac{1}{10}$, find an approximation for $\sqrt{10}$ in the form $\frac{a}{b}$, where a and b are to be determined. [2]

4 Trigonometry

Essential facts and formulae

Facts

Showing by counter example – to prove a given statement is false you need to find one case for which the statement is not true.

Formulae

Radian measure, arc length, area of sector and area of segment

$$\pi \text{ radians} = 180° \qquad\qquad 2\pi \text{ radians} = 360°$$

$$\frac{\pi}{2} \text{ radians} = 90° \qquad\qquad \frac{\pi}{4} \text{ radians} = 45°$$

$$\frac{\pi}{3} \text{ radians} = 60° \qquad\qquad \frac{\pi}{6} \text{ radians} = 30°$$

The length of an arc making an angle of θ radians at the centre $l = r\theta$

Area of sector making an angle of θ radians at the centre $= \frac{1}{2}r^2\theta$

Area of segment $= \frac{1}{2}r^2(\theta - \sin\theta)$

sec, cosec and cot

$$\sec\theta = \frac{1}{\cos\theta} \qquad\qquad \operatorname{cosec}\theta = \frac{1}{\sin\theta}$$

$$\cot\theta = \frac{1}{\tan\theta}$$

Trignometric identities

$$\sec^2\theta = 1 + \tan^2\theta$$

$$\operatorname{cosec}^2\theta = 1 + \cot^2\theta$$

$$\sin(A \pm B) = \sin A \cos B \pm \cos A \sin B$$

$$\cos(A \pm B) = \cos A \cos B \mp \sin A \sin B$$

$$\tan(A \pm B) = \frac{\tan A \pm \tan B}{1 \mp \tan A \tan B}$$

Double angle formulae

$$\sin 2A = 2\sin A \cos A$$

$$\cos 2A = \cos^2 A - \sin^2 A$$

$$= 1 - 2\sin^2 A$$

$$= 2\cos^2 A - 1$$

$$\tan 2A = \frac{2\tan A}{1 - \tan^2 A}$$

Important rearrangements of the double angle formulae

$$\sin^2 A = \frac{1}{2}\left(1 - \cos 2A\right)$$

$$\cos^2 A = \frac{1}{2}\left(1 + \cos 2A\right)$$

Use of small angle approximation for sine, cosine and tangent

If the angle θ is small and measured in radians:

$$\sin \theta \approx \theta$$

$$\cos \theta \approx 1 - \frac{\theta^2}{2}$$

$$\tan \theta \approx \theta$$

Questions

1 Prove that $\dfrac{\sin (A - B)}{\cos A \cos B} = \tan A - \tan B$ [2]

2 (a) Express $3\cos\theta + 4\sin\theta$ in the form $R\cos(\theta - \alpha)$, where $R > 0$ and $0° < \alpha < 90°$. [3]

(b) Use your results to part (a) to find the least value of

$$\frac{1}{3\cos\theta + 4\sin\theta + 7}$$

Write down a value of θ for which this least value occurs. [2]

3 Show by counter-example, that the statement:

'If $\cos\theta = \cos\phi$ then $\sin\theta = \sin\phi$'

is false. [3]

(a) On the same diagram, sketch the graphs of $y = \cos^{-1} x$ and $y = 5x - 1$. [2]

(b) Prove that the equation:

$$\cos^{-1} x - 5x + 1 = 0$$

has a root α between 0.4 and 0.5. [3]

5 Given that x is a small angle, show that:

$$\cos(x - \alpha) - \cos \alpha = x \sin \alpha - \frac{x^2}{2} \cos \alpha$$ [3]

6 For all values of x, $f(x) = \sin x + \cos x + 3$

   (a) Find the maximum and minimum values of $f(x)$.   [5]

   (b) Find the maximum and minimum values of $\dfrac{1}{f(x)}$.   [2]

7 Given that, $2\sin(x - 60°) = \cos(\theta + 60°)$

   show that $\tan x = a\sqrt{3} + b$ where a and b are integers.   [6]

5 Differentiation

Essential facts and formulae

Facts

Finding points of inflection

To find points of inflection, find the second derivative and equate the resulting equation to zero. Solve the equation and put the x-value or values back into the equation of the curve to determine the y-coordinate(s). Check the sign of the second derivative either side of each x-coordinate to check that there is a sign change. If there is, then the point is a point of inflection.

Formulae

Differentiation of e^{kx}, a^{kx}, $\sin kx$, $\cos kx$ and $\tan kx$

$$\frac{d(e^{kx})}{dx} = ke^{kx}$$

$$\frac{d(a^{kx})}{dx} = ka^{kx}\ln a$$

$$\frac{d(\sin kx)}{dx} = k\cos kx$$

$$\frac{d(\cos kx)}{dx} = -k\sin kx$$

> All these derivatives need to be remembered.

The following derivative need not be remembered as it is included in the formula booklet.

$$\frac{d(\tan kx)}{dx} = k\sec^2 kx$$

The derivative of $\ln x$

$$\frac{d(\ln x)}{dx} = \frac{1}{x}$$

> This derivative needs to be remembered.

To differentiate the natural logarithm of a function you differentiate the function and then divide by the function.

This can be expressed mathematically as:

$$\frac{d(\ln(f(x))}{dx} = \frac{f'(x)}{f(x)}$$

The Chain rule

If y is a function of u and u is a function of x, then the chain rule states:

$$\frac{dy}{dx} = \frac{dy}{du} \times \frac{du}{dx}$$

The Product rule

If $y = f(x)\,g(x)$, $\qquad \dfrac{dy}{dx} = f(x)\,g'(x) + g(x)\,f'(x)$

The Quotient rule

If $y = \dfrac{f(x)}{g(x)}$, $\qquad \dfrac{dy}{dx} = \dfrac{f'(x)\,g(x) - f(x)\,g'(x)}{(g(x))^2}$

Differentiation of inverse functions $\sin^{-1} x$, $\cos^{-1} x$, $\tan^{-1} x$

$$\frac{d(\sin^{-1} x)}{dx} = \frac{1}{\sqrt{1 - x^2}}$$

$$\frac{d(\cos^{-1} x)}{dx} = -\frac{1}{\sqrt{1 - x^2}}$$

$$\frac{d(\tan^{-1} x)}{dx} = -\frac{1}{1 + x^2}$$

Differentiation of simple functions defined implicitly

Finding $\dfrac{dy}{dx}$ in terms of both x and y is called implicit differentiation.

Here are the rules for differentiating implicitly:

- Terms involving x or constant terms are differentiated as normal.

- For terms just involving y, (e.g. $3y$, $5y^3$, etc.) differentiate with respect to y and then multiply the result by $\frac{dy}{dx}$.

- For terms involving both x and y (e.g. xy, $5x^2y^3$, etc.) the Product rule is used because there are two terms multiplied together. Note the need to include $\frac{dy}{dx}$ when the term involving y is differentiated.

Differentiation of functions defined parametrically

The equation of a curve can be expressed in parametric form by using:

$$x = f(t), y = g(t) \qquad \text{where } t \text{ is the parameter being used.}$$

The formulae for differentiating parametric forms are:

$$\frac{dy}{dx} = \frac{\frac{dy}{dt}}{\frac{dx}{dt}} = \frac{dy}{dt} \times \frac{dt}{dx}$$

and

$$\frac{d^2y}{dx^2} = \frac{\frac{d}{dt}\left(\frac{dy}{dx}\right)}{\frac{dx}{dt}} = \frac{d}{dt}\left(\frac{dy}{dx}\right) \times \frac{dt}{dx}$$

Questions

1 A spherical balloon is inflated at the rate of 0.05 m³ per minute.
Find the rate of increase of surface area when the radius is 8 cm. [4]

2 Differentiate each of the following with respect to x, simplifying your answer where possible.

(a) $(3x^2 - 2x)^7$ [2]

(b) $\sqrt{3x^3 - 4}$ [2]

(c) $x^3 \ln 3x$ [2]

(d) $\dfrac{e^{2x}}{(2x + 1)^5}$ [3]

3 Differentiate each of the following with respect to x, simplifying your answer where possible.

(a) $\ln(9x^3 - 2x + 1)$ [2]

(b) $e^{\sqrt{x}}$ [2]

(c) $\dfrac{a - b\cos x}{a + b\sin x}$, where a and b are constants. [2]

4 The function f is defined by:

$$f(x) = \frac{6 + x - 9x^2}{x^2(x + 2)}$$

Verify that $f(x)$ has a stationary value when $x = 2$. [7]

5 Find the equation of the tangent to the curve $x^3 - 2xy^2 + y^3 = 5$
at the point $(2, 1)$. [5]

6 The curve C has parametric equations $x = 2t$, $y = 5t^3$.
Show that the equation of the tangent C at the point P is:
$$2y = 15p^2x - 20p^3$$ [5]

7 Differentiate each of the following with respect to x, simplifying your answer wherever possible.

(a) $\ln(\cos x)$ [2]

(b) $\tan^{-1}\left(\frac{x}{2}\right)$ [2]

(c) $e^{4x}(2x-1)^4$ [2]

8 Differentiate each of the following with respect to x, simplifying your answer wherever possible.

(a) $\tan^{-1}(5x)$ [2]

(b) e^{x^2} [2]

(c) $x^3 \ln x$ [2]

(d) $\dfrac{5-x^2}{3x^2-1}$ [2]

9 Given that $x = 2t - \sin 2t$, $y = \cos 3t$, show that $\dfrac{dy}{dx} = \dfrac{3}{4}$ when $t = \dfrac{\pi}{2}$. [5]

10 The function f is defined by $f(x) = x^2 e^x$.

(a) Show that $f'(x) = xe^x(x + 2)$. [2]

(b) Find the value of $f'(x)$ in terms of e when $x = 1$. [1]

6 Coordinate geometry in the (x, y) plane

Essential facts and formulae

Facts

Cartesian and parametric equations

Cartesian equations connect x and y in some way. For example, $y = 4x^3$ is a Cartesian equation.

Parametric equations express x and y in terms of a parameter such as t, for example

$$x = 4 + 2t \qquad y = 1 + 2t$$

To obtain the Cartesian equation from the parametric equation it is necessary to eliminate the parameter.

Note that $\dfrac{dy}{dx} = \dfrac{dy}{dt} \times \dfrac{dt}{dx}$

Using the Chain rule to find the second derivative

The Chain rule can be used to find the second derivative in terms of a parameter such as t in the following way:

$$\frac{d^2y}{dx^2} = \frac{d}{dx}\left(\frac{dy}{dx}\right) = \frac{d}{dt}\left(\frac{dy}{dx}\right)\frac{dt}{dx}$$

Implicit differentiation

Here are the basic rules:

$$\frac{d(3x^2)}{dx} = 6x$$

> Terms involving x or constant terms are differentiated as normal.

$$\frac{d(6y^3)}{dx} = 18y^2 \times \frac{dy}{dx}$$

> Differentiate with respect to y and then multiply the result by $\frac{dy}{dx}$.

$$\frac{d(y)}{dx} = 1 \times \frac{dy}{dx}$$

> When you are differentiating a term just involving y, you differentiate with respect to y and then multiply the result by $\frac{dy}{dx}$.
> This is an application of the Chain rule.

$$\frac{d(x^2y^3)}{dx} = (x^2)\left(3y^2 \times \frac{dy}{dx}\right) + (y^3)(2x) = 3x^2y^2\frac{dy}{dx} + 2xy^3$$

> Because there are two terms here, the Product rule is used.
> Notice the need to include $\frac{dy}{dx}$ when the term involving y is differentiated.

33

Questions

1 Find the equation of the normal to the curve
$$x^2 + 3xy + 3y^2 + 13$$
at the point (2, 1). [7]

2 The parametric equations of the curve C are $x = t^2, y = t^3$.
The point P has parameter p.

(a) Show that the equation of the tangent to C at the point P is $3px - 2y = p^3$ [4]

(b) The tangent to C at the point P intersects C again at the point $Q(q^2, q^3)$.
Given that $p = 2$, show that q satisfies the equation $q^3 - 3q^2 + 4q = 0$
and determine the value of q. [5]

3 A function is defined parametrically by:
$$x = 4 \sin 2t, \quad y = 3 \cos 2t.$$

(a) Find and simplify an expression for $\dfrac{d^2y}{dx^2}$. [3]

(b) Find and simplify an expression for $\dfrac{d^2y}{dx^2}$
  (i) in terms of t,
  (ii) in terms of y. [4]

4 (a) The curve C is given by the equation:
$$x^4 + x^2y + y^2 = 13$$
  Find the value of $\dfrac{dy}{dx}$ at the point $(-1, 3)$. [4]

(b) Show that the equation of the normal to the curve $y^2 = 4x$, at the point $P(p^2, 2p)$ is:
$$y + px = 2p + p^3$$
  Given that $p \neq 0$ and that the normal at P cuts the x-axis at $B(b, 0)$, show that $b > 2$. [5]

5 The curve C has parametric equations $x = t^2 - 4$, $y = 3t^4 + 8t^3$.

Find the equation of the tangent to C at the point where $t = -1$. [5]

7 Integration

Essential facts and formulae

Facts

If you are asked to find the area bounded by a curve – sketch the curve first to check that there is not part of the curve below the x-axis that would have a negative area.

Formulae

Integration of $x^n(n \neq -1)$, e^{kx}, $\frac{1}{x}$, $\sin kx$, $\cos kx$

$$\int x^n \, dx = \frac{x^{n+1}}{n+1} + c$$

$$\int e^{kx} \, dx = \frac{e^{kx}}{k} + c$$

$$\int \frac{1}{x} \, dx = \ln|x| + c$$

$$\int \sin kx \, dx = -\frac{1}{k} \cos kx + c$$

$$\int \cos kx \, dx = \frac{1}{k} \sin kx + c$$

> You will be required to remember these results as they are not given in the formula booklet.

Integration of $(ax+b)^n \, (n \neq -1)$, e^{ax+b}, $\frac{1}{ax+b}$, $\sin(ax+b)$, $\cos(ax+b)$

$$\int (ax+b)^n \, dx = \frac{(ax+b)^{n+1}}{(n+1)a} + c \qquad (n \neq -1)$$

$$\int e^{ax+b} \, dx = \frac{e^{ax+b}}{a} + c$$

$$\int \frac{1}{ax+b} \, dx = \frac{1}{a} \ln|ax+b| + c$$

$$\int \sin(ax+b) \, dx = \frac{-\cos(ax+b)}{a} + c$$

$$\int \cos(ax+b) \, dx = \frac{\sin(ax+b)}{a} + c$$

> You will be required to remember these results as they are not given in the formula booklet.

Integration by substitution and integration by parts

An integral of the type $\int f(x)\, dx$ is converted into the integral $\int f(x)\frac{dx}{du}\, du$, where x is replaced by a given substitution. In the case of definite integrals, the x limits are converted into u limits by means of the given substitution.

Integration by parts

Integration by parts is used when there is a product to integrate.

$$\int u \frac{dv}{dx}\, dx = uv - \int v \frac{du}{dx}\, dx$$

Integration using partial fractions

Single fractions such as $\frac{5x+3}{(x+3)(x-1)}$ can be converted into partial fractions $\left(\frac{3}{(x+3)} + \frac{2}{(x-1)} \text{ in this case}\right)$ so that each of the resulting partial fractions can be integrated.

In many cases the answer to these questions involves the use of ln.

Analytical solution of first order differential equations with separable variables

Equations of the type

$$\frac{dy}{dx} = f(x)g(y)$$

can be solved by separating the variables and integrating both sides of the resulting equation, i.e.

$$\int \frac{1}{g(y)}\, dy = \int f(x)\, dx$$

Questions

1 Show that $= \int_0^1 \frac{e^x}{1+e^x}\, dx = \ln\frac{1+e}{2}$ [5]

2 Find each of the following, simplifying your answer wherever possible.

(a) $\int e^{\frac{x}{2}}\, dx$ [2]

(b) $\int \frac{5x}{5x^2+9}\, dx$ [3]

3 Find $\displaystyle\int \frac{x}{(2x+1)^3}\,dx$ [5]

4 Evaluate $\displaystyle\int_{-\frac{1}{4}}^{\frac{1}{4}} 2x(4x+1)^6\,dx$ [6]

5 Find the value of $\int_0^2 \dfrac{6}{(2x-3)^2}\,dx$ [5]

6 The graph below shows the curves $y = \sin x$ and $y = \frac{1}{2}\sin 2x$.
Also shown is the line $x = k$ where k is a constant.

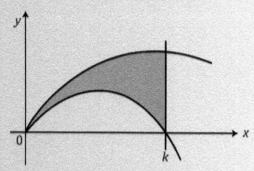

(a) Find the value of k. [3]
(b) Calculate the area of the shaded region. [5]

7 Show that $\displaystyle\int_3^4 \frac{4x+1}{(x+1)^2(x-2)}\,dx = \frac{1}{20} + \ln\frac{8}{5}$ [5]

8 Find the value of $\displaystyle\int_1^2 \frac{6}{(2x-3)^3}\,dx$ [4]

9 Given that
$$f(x) = \frac{3x}{(1+x)^2(2+x)}$$

(a) Express $f(x)$ in terms of partial fractions. [4]

(b) Evaluate:

$$\int_0^1 f(x)\, dx,$$

Giving your answer correct to three decimal places. [4]

10 Find $\int x\sqrt{1+x^2}\, dx$ [4]

11 Find $\int \frac{x^3}{1 + x^4} \, dx$ [4]

12 Find $\int \frac{\cos x}{\sin^3 x} \, dx$ [5]

13 Find $\int \frac{5}{2 + 3x} \, dx$ [3]

(a) Integrate:

(i) e^{-3x+5} [2]

(ii) $x^2 \ln x$ [4]

(b) Use an appropriate substitution to show that:

$$\int_0^{\frac{1}{2}} \frac{x^2}{\sqrt{1-x^2}}\, dx = \frac{\pi}{12} - \frac{\sqrt{3}}{8}$$ [8]

15 Find:

(a) $\int \frac{1}{1-x}\, dx$ [1]

(b) $\int (2x-3)^5\, dx$ [2]

(c) $\int 5 \sin(2x-1)\, dx$ [2]

8 Numerical methods

Essential facts and formulae

Facts

Location of roots of $f(x) = 0$, considering changes of sign of $f(x)$

If $f(x)$ can take any value between a and b, then if there is a change of sign between $f(a)$ and $f(b)$, then a root of $f(x)$ lies between a and b.

Formulae

Newton–Raphson iteration

Newton–Raphson iteration for solving $f(x) = 0$

$$x_{n+1} = x_n - \frac{f(x_n)}{f'(x_n)}$$

The Trapezium rule for estimating the area under a curve or the integral of a function

The Trapezium rule can be used for estimating areas or working out definite integrals of functions where the function is too difficult to integrate.

$$\int_a^b y \, dx \approx \frac{1}{2} h\{(y_0 + y_n) + 2(y_1 + y_2 + \ldots + y_{n-1})\} \quad \text{where } h = \frac{b-a}{n}$$

Questions

 Use the Trapezium rule with five ordinates to find an approximate value for the integral:

$$\int_{4}^{6} \frac{1}{3 - \sqrt{x}}\,dx$$

Show your working and give your answer correct to three decimal places.   [4]

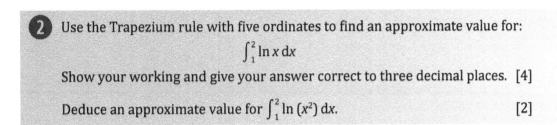

2 Use the Trapezium rule with five ordinates to find an approximate value for:

$$\int_1^2 \ln x \, dx$$

Show your working and give your answer correct to three decimal places. [4]

Deduce an approximate value for $\int_1^2 \ln (x^2) \, dx$. [2]

3 (a) Sketch the graphs of $y = x^2$ and $y = \cos x$ on the same set of axes in the interval $0 \le x \le \frac{1}{2}\pi$. [2]

(b) Show that the root of the equation $x^2 = \cos x$ lies in the interval $0.8 \le x \le 0.9$. [2]

(c) Use the Newton–Raphson method with a starting value of 0.8 to find an estimate of this root, giving your answer to three decimal places. [5]

(a) On the same diagram, sketch the graphs of $y = \cos^{-1} x$ and $y = 5x - 1$. [2]

(b) You may assume that the equation $\cos^{-1} x - 5x + 1 = 0$ has a root α between 0.4 and 0.5.

The recurrence relation:

$$x_{n+1} = \frac{1}{5}\left(1 + \cos^{-1} x_n\right)$$

with $x_0 = 0.4$ can be used to find α.

Find and record the values of x_1, x_2, x_3, x_4. Write down the value of x_4 correct to four decimal places and prove that this is the value of α correct to four decimal places. [5]

1 Probability

Essential facts and formulae

Facts

Independent events – events where the probability of a first event occurring does not influence the probability of a second event occurring

Conditional probability – the probability of an event occurring given that another event has occurred

Mutually exclusive events – event A can happen or event B can happen but not both

Formulae

The multiplication law for independent events

If events A and B are independent:

$$P(A \cap B) = P(A) \times P(B)$$

The As and Bs can be swapped around in this formula and you can also swap A for A' and B for B' for all combinations.

The multiplication law for dependent events

If events A and B are dependent:

$$P(A \cap B) = P(A) \times P(B|A)$$

The generalised addition law

$$P(A \cap B) = P(A) + P(B) - P(A \cup B)$$

The generalised addition law can be used for dependent or independent events (A').

This formula is included in the formula booklet.

The conditional probability formula

$P(A|B)$ means the probability of A given that B has occurred.

$$P(A|B) = \frac{P(A \cap B)}{P(B)}$$

The As and Bs can be swapped around in this formula and you can also swap A for A' and B for B' for all combinations.

Questions

1. There are 3 red and 7 blue counters in a bag and two counters are picked at random from the bag.

   Find the probability that:

   (a) Two red counters are chosen. [1]

   (b) A red and blue counter are chosen. [2]

2. Two independent events A and B are such that:
   $$P(A) = 0.4, \quad P(B) = 0.3$$

   Find:

   (a) $P(A \cap B)$, [1]

   (b) $P(A \cup B)$, [3]

   (c) the probability that neither A nor B occurs, [3]

   (d) $P(A|A \cup B)$. [3]

3 The events A and B are such that $P(A) = 0.3$, $P(B) = 0.4$, $P(A \cup B) = 0.5$.

(a) Determine whether or not A and B are independent. [3]

(b) Evaluate $P(A|B')$. [3]

4 The events A and B are such that $P(A) = 0.5$, $P(B) = 0.3$.

(a) Evaluate $P(A \cup B)$ when:

(i) A, B are mutually exclusive,

(ii) A, B are independent. [5]

(b) Given that $P(A \cup B) = 0.7$, find the value of $P(B|A)$. [3]

5 Using the two-way table shown here, find each of the following:

(a) P(C') [1]

(b) P($D' \cap C'$) [2]

(c) P($D \cup C$) [2]

	D	D'	Total
C	12	40	52
C'	32	16	48
Total	44	56	100

6 (a) Marie is an athlete who competes in the high jump. In a certain competition she is allowed two attempts to clear each height, but if she is successful with the first attempt she does not jump again at this height. The probability that she is successful with her first jump at a height of 1.7 m is p. The probability that she is successful with her second jump is also p. The probability that she clears 1.7 m is 0.64. Find the value of p. [4]

(b) The table shows the numbers of male and female athletes competing for Wales in track and field events at a competition:

	Track	Field
Male	13	9
Female	7	4

Two athletes are chosen at random to participate in a drugs test. Given that the first athlete is male, find the probability that both are field athletes. [3]

7 An architect bids for two construction projects. He estimates the probability of winning bid A is 0.6, the probability of winning bid B is 0.5 and the probability of winning both is 0.2 .

(a) Show that the probability that he does not win either bid is 0.1. [2]

(b) Find the probability that he wins exactly one bid. [2]

(c) Given that he does not win bid A, find the probability that he wins bid B. [3]

2 Statistical distributions

Essential facts and formulae

Facts

Continuous uniform distribution – a probability distribution where all values in the same interval of allowed values, have equal probabilities of occurring

Normal distribution – a continuous distribution that enables you to find the probability of a quantity taking certain values. The graph of the distribution is a bell-shaped curve which is symmetrical either side of the mean value

Formulae

Continuous uniform distribution

If $X \sim \text{U}[a, b]$, then:

$$\text{Mean, } \text{E}(X) = \frac{1}{2}\left(a + b\right)$$

$$\text{Variance, } \text{Var}(X) = \frac{1}{12}\left(b - a\right)^2$$

$$\text{P}(c \leq X \leq d) = \frac{d - c}{b - a}$$

> Both of these formulae are included in the formula booklet.

> This last formula is not included in the formula booklet and will need to be remembered.

Normal distribution

If $X \sim \text{N}(\mu, \sigma^2)$
the distribution is:

For the standard normal distribution, the distribution is adjusted to:

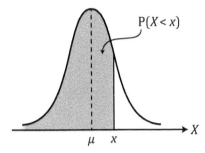

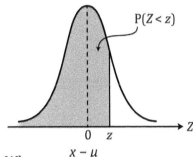

$$\text{Where } z = \frac{x - \mu}{\sigma}$$

The normal distribution function table or a calculator is used to find $\text{P}(Z < z)$

Questions

 Bags of sugar are labelled as having a weight of 1 kg. However, their actual weight has a continuous uniform distribution over the interval 0.980 kg to 1.030 kg.

Find the probability that the net weight of a bag picked at random is less than 1 kg. [2]

2 Antonio arrives at a train station at a random point in time. The trains to his desired destination are scheduled to depart at 12-minute intervals.

(a) Assume that Antonio gets on the next train.

   (i) Suggest an appropriate distribution to model his waiting time and give the parameters.

   (ii) State the mean and the variance of this distribution.

   (iii) State an assumption you have made in suggesting this distribution. [4]

(b) Now assume that the probability that Antonio misses the next available train because he is distracted by his smartphone is 0.12. If he misses the next available train, he is sure to get on the one after that.

   (i) Find the probability that he waits between 9 and 19 minutes.

   (ii) Given that he waits between 9 and 19 minutes, find the probability that he gets on the first train. [6]

3 A continuous random variable X has a continuous uniform distribution over the interval −6 to 4.

   (a) Write down the mean of X. [2]

   (b) Find $P(X \leq 2.4)$. [2]

   (c) Find $P(-2 \leq X \leq 3)$. [2]

   (d) Find $P(-8 \leq X - 3 \leq 3)$. [3]

The volume of ice cream, X litres, in a catering tub of ice cream is normally distributed with mean 5 litres and variance σ^2.

(a) If it is assumed that the variance is 0.09, find $P(X < 4.5)$. [3]

(b) Find the value of σ so that 95% of the tubs contain more than 4.8 litres of ice cream. [4]

 X is a normally distributed random variable with a mean of 15 and a standard deviation of 4.

Find the value of x, correct to two decimal places such that:

(a) $P(X < x) = 0.45$ [2]

(b) $P(X > x) = 0.3$ [2]

(c) The variable X was investigated by someone else and they found that the mean had changed to 20 and the standard deviation had changed to 3. Find the value of x, correct to two decimal places such that:

$$P(18 < X < x) = 0.4$$ [3]

6 Arwyn collects data about household expenditure on food. He records the weekly expenditure on food for 80 randomly selected households from across Wales.

Cost, x (£)	$x < 40$	$40 \leq x < 50$	$50 \leq x < 60$	$60 \leq x < 70$	$70 \leq x < 80$	$80 \leq x < 90$	$x \geq 90$
Number of households	5	11	16	18	15	9	6

(a) Explain why a Normal distribution may be an appropriate model for the weekly expenditure on food for this sample. [1]

Arwyn uses the distribution $N(64, 15^2)$ to model expenditure on food.

(b) Find the number of households in the sample that this model would predict to have weekly food expenditure in the range:

   (i) $60 \leq x \leq 70$,

   (ii) $x \geq 90$. [4]

(c) Use your answers to part (b):

   (i) to comment on the suitability of this model,

   (ii) to explain how Arwyn could improve the model by changing one of its parameters. [2]

(d) Arwyn's friend, Colleen, wishes to use the improved model to predict household expenditure on food in Northern Ireland. Comment on this plan. [1]

7 X is a random variable that can be modelled using a Normal distribution having mean μ and standard deviation σ.

(a) If $P(X < 2) = 0.2$ and $P(X > 4) = 0.15$, find the value of μ and σ giving your answers correct to 3 decimal places. [7]

(b) Find $P(X > 3)$ giving your answer correct to 3 decimal places. [3]

3 Statistical hypothesis testing

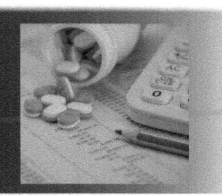

Essential facts and formulae

Facts

Correlation coefficients

Have the letters ρ if based on a population or r if based on a sample.

They have values between and including -1 and 1.

   ρ or $r = 0$ means no correlation

   ρ or $r = 1$ means perfect positive correlation

   ρ or $r = -1$ means perfect negative correlation

Performing hypothesis testing using a correlation coefficient as a test statistic

Testing for positive correlation use $\mathbf{H_0} : \rho = 0$, $\mathbf{H_1} : \rho > 0$

Testing for negative correlation use $\mathbf{H_0} : \rho = 0$, $\mathbf{H_1} : \rho < 0$

Testing for any correlation (i.e. positive or negative) use $\mathbf{H_0} : \rho = 0$, $\mathbf{H_1} : \rho \neq 0$

> Note ρ can be replaced with r in all of these tests.

Hypothesis testing for the mean of a normal distribution with a known, given or assumed variance

If X is normally distributed then $X \sim N(\mu, \sigma^2)$ and the sample mean, $\overline{X}$, is normally distributed, so $\overline{X} \sim N\left(\mu, \frac{\sigma^2}{n}\right)$

If $Z \sim N(0, 1)$, then

$$Z = \frac{\overline{X} - \mu}{\frac{\sigma}{\sqrt{n}}}$$

> This formula is used to work out z-values that apply to the standard normal distribution.

Hence the z-value (i.e. the test statistic) will be:

$$z = \frac{\overline{X} - \mu}{\frac{\sigma}{\sqrt{n}}}$$

There are two methods that can be used for hypothesis testing:

- Using the z-value to find the probability (i.e. the p-value) and then compare this probability to the significance level.

- Use the significance level of the test to find the critical value and then see if the test statistic (i.e. the sample mean) lies inside or outside the critical region.

Questions

 A chain of coffee bars is looking to purchase a large number of new coffee machines.

At present, the times in seconds to produce a standard Americano coffee from beans follows a N(25, 4) distribution. The manufacturer loans the coffee bars' owner a new machine to test and in a random sample of 20 coffees, they took a mean time of 24 seconds to produce. It is assumed that the standard deviation of the sample has remained the same.
Test at the 5% level of significance whether the mean time in producing coffee has decreased with the new machine. [4]

2 Rebecca is a farmer who is monitoring prices for products to use on her farm. She records the prices of two products made from **different** grains, wheat and oats, at random points in time, to investigate whether there is any correlation.

Price of feed wheat versus price of feed oats

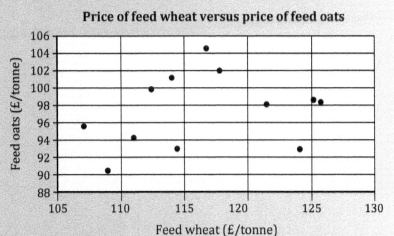

The product moment correlation coefficient for the data is 0.244. There are 12 data points, and the p-value is 0.4447.

(a) Comment on the correlation between the prices of feed wheat and feed oats.   [2]

Rebecca also records the prices of two wheat products at random points in time, to investigate whether there is any correlation.

Price of feed wheat versus price of wheat straw

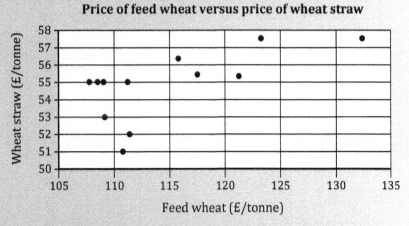

The product moment correlation coefficient for the data is 0.653. There are 12 data points.

(b) Stating your hypotheses clearly, test at the 5% level of significance whether there is any evidence of correlation between the prices of these two products.   [5]

(c) Without referring to the positioning of the points on the graphs, suggest why the product moment correlation coefficient is higher for the second set of data.   [1]

Your answer to Question 2

A fitness trainer is investigating the BMI (Body Mass Index) and the % fat people have in their body to see if there is any correlation between them. A random sample of 30 members of the gym was taken, and their BMI and % fat measured.

The scatter graph shows the relationship between 'BMI' and '% fat' for this sample of gym members.

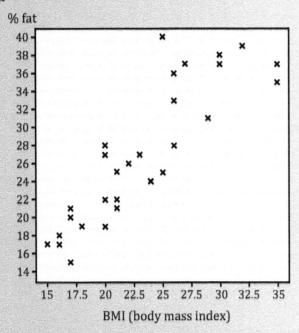

(a) The product moment correlation coefficient for 'BMI' and '% Fat' is 0.8670.

Stating your hypothesis clearly, test, at the 5% level of significance, whether this correlation is significant. State your conclusion in context. [4]

(b) The fitness trainer also investigates if % Fat has any significant correlation with daily calories consumed. She used statistical analysis software to produce the following output. The output in the following table shows the Pearson correlation coefficient and the corresponding *p*-value for the two quantities of interest.

	% Fat	Daily calories
% Fat	1	
Daily calories	*p*-value: 0.086 Correlation coefficient: 0.78	1

State with a reason whether there is any correlation between 'BMI' and '% fat'. [2]

Your answer to Question 3 continued

A machine in a factory produces screws with a mean length of 1.5 cm. The lengths are normally distributed and have a standard deviation of 0.05 cm.

The machine needs to be periodically adjusted so that it is making screws with the mean length of 1.5 cm.

To find out if it needs adjusting, a sample of 50 screws is taken and their lengths measured. It is found that the mean length has changed. It is assumed that the mean lengths of the screws in the sample follow a Normal distribution and have the same standard deviation of 0.05 cm.

(a) Stating your hypotheses clearly, find at the 1% level of significance, the critical region for this test. [6]

(b) The mean length of the sample of screws was found to be 1.53 cm. State, giving a reason, whether the machine producing the screws needs adjusting or not. [2]

4 Kinematics for motion with variable acceleration

Essential facts and formulae

Facts

Using calculus when the acceleration is not constant

Calculus has to be used to find s, v, or a when the acceleration is not constant, and the following diagrams show the processes involved.

Using differentiation

Displacement (s) $\xrightarrow[\dfrac{ds}{dt}]{\text{Differentiate}}$ Velocity (v) $\xrightarrow[\dfrac{dv}{dt}]{\text{Differentiate}}$ Acceleration (a)

Using integration

Displacement (s) $\xleftarrow[s = \int v\,dt]{\text{Integrate}}$ Velocity (v) $\xleftarrow[v = \int a\,dt]{\text{Integrate}}$ Acceleration (a)

Formulae

$$v = \frac{ds}{dt}$$

$$a = \frac{dv}{dt}$$

$$s = \int v\,dt$$

$$v = \int a\,dt$$

Newton's 2nd law of motion

A resultant force, F N, produces acceleration, a m s^{-2}, on a mass, m kg, according to the formula

Force = mass × acceleration or $F = ma$.

Questions

 A car travels in a straight line and t seconds after passing a point P it has a velocity given by $v = 64 - \frac{1}{27}t^3$.

The car comes to rest at point Q.

(a) Show that the car comes to rest after $t = 12$ s. [3]

(b) Calculate the distance PQ. [2]

2 The initial speed of a particle is 10 m s^{-1} after passing a point A and it is subjected to an acceleration of $(2t + 3) \text{ m s}^{-2}$.

Calculate:

(a) the speed of the particle after 3 seconds, [3]

(b) the distance the particle travels 3 seconds after passing point A. [4]

3 A particle moves along the x-axis so that its velocity at time t seconds is $v\,\text{m s}^{-1}$, where $v = 4t^3 - 6t + 1$. Given that the displacement of P is 4 m from the origin when $t = 0$, find:

(a) the displacement of P from the origin when $t = 2$, [4]

(b) the acceleration of P when $t = 1$. [2]

4 A particle moves from point O in a straight line and its velocity $v\,\text{m s}^{-1}$ at time t seconds is given by

$$v = 2t^2 - 20t + 32 \qquad \text{for } t \geq 0$$

(a) Find the two times when the velocity is instantaneously at rest and hence draw a sketch of the velocity–time graph. [3]

(b) Determine the minimum value of the particle's velocity. [3]

(c) Find the values of t when the velocity of the particle is $14\,\text{m s}^{-1}$. [2]

(d) Find the total distance travelled between $t = 0\,\text{s}$ and $t = 8\,\text{s}$. [3]

5 Kinematics for motion using vectors

Essential facts and formulae

Facts

A **scalar** quantity has magnitude (i.e. size) only – examples include distance, speed and time.

A **vector** quantity has both magnitude and direction – examples include displacement, velocity, acceleration and force.

Formulae for constant acceleration for motion in a straight line using vectors

$$\mathbf{v} = \mathbf{u} + \mathbf{a}t \qquad\qquad \mathbf{v}^2 = \mathbf{u}^2 + 2\mathbf{a}\mathbf{s}$$

$$\mathbf{s} = \mathbf{u}t + \frac{1}{2}\mathbf{a}t^2 \qquad\qquad \mathbf{s} = \frac{1}{2}(\mathbf{u} + \mathbf{v})t$$

$\mathbf{s}$ = displacement vector
$\mathbf{u}$ = initial velocity vector
$\mathbf{v}$ = final velocity vector
$\mathbf{a}$ = acceleration vector
t = time

Newton's 2nd law of motion

Force = mass × acceleration or $\mathbf{F} = m\mathbf{a}$

Using calculus for motion in a straight line using vectors

For questions involving variable acceleration we make use of the following equations:

$$\mathbf{v} = \frac{d\mathbf{s}}{dt} \qquad\qquad \mathbf{a} = \frac{d\mathbf{v}}{dt}$$

$$\mathbf{s} = \int \mathbf{v}\, dt \qquad\qquad \mathbf{v} = \int \mathbf{a}\, dt$$

Finding the magnitude of a vector

The magnitude of displacement is distance (a scalar).

The magnitude of velocity is speed (a scalar).

The magnitude of acceleration or force does not have its own name.

For a vector in two dimensions such as vector, $\mathbf{v} = a\mathbf{i} + b\mathbf{j}$,
then scalar/magnitude of vector, $|\mathbf{v}| = \sqrt{a^2 + b^2}$

For a vector in three dimensions such as vector, $\mathbf{v} = a\mathbf{i} + b\mathbf{j} + c\mathbf{k}$
then scalar/magnitude of vector, $= |\mathbf{v}| = \sqrt{a^2 + b^2 + c^2}$

Questions

1 A particle of mass 2 kg moves in a straight line on a horizontal plane under the action of the following three forces; $2\mathbf{i} + \mathbf{j}$, $-\mathbf{i} + 2\mathbf{j}$ and $3\mathbf{i} - \mathbf{j}$.

    (a) Find the vector, $\mathbf{F}$, of the resultant force. [2]

    (b) The velocity of the particle at time t s is $\mathbf{v}$ m s^{-1}. When $t = 0$ s, the velocity is $(3\mathbf{i} - 7\mathbf{j})$ m s^{-1}. Find an expression for $\mathbf{v}$ in terms of t. [5]

    (c) Find the speed after 2 s giving your answer to 1 decimal place. [2]

2 An object of mass 5 kg is moving on a horizontal plane under the action of a constant force $5\mathbf{i} + 15\mathbf{j}$ N. At time $t = 0$ s, its position vector is $3\mathbf{i} - 12\mathbf{j}$ with respect to the origin O and its velocity vector is $-3\mathbf{i} + 5\mathbf{j}$.

    (a) Determine the velocity vector of the object at time $t = 2$ s. [3]

    (b) Calculate the exact value of the distance of the object from the origin when $t = 2$ s. [5]

3 A child's toy robot starts from a point on a horizontal floor with a position vector $2\mathbf{i} + 4\mathbf{j}$ with respect to a fixed origin O with a fixed speed of $1.3 \, \text{m s}^{-1}$ and travels towards another point on the floor with a position vector $7\mathbf{i} + 16\mathbf{j}$. Find the time taken for the robot to travel between these two points. [5]

4 A particle moves in a horizontal plane with a velocity $\mathbf{v} \, \text{m s}^{-1}$ at time $t \, \text{s}$ where $\mathbf{v}$ is given by:

$$\mathbf{v} = 2 \tan 2t\mathbf{i} \quad \text{for } 0 \le t \le \frac{\pi}{3}$$

Calculate the acceleration when $t = \frac{\pi}{6} \, \text{s}$. [4]

5 The position vector of a particle P at time t seconds is given by

$$\mathbf{r} = t \sin t\mathbf{i} + t \cos t\mathbf{j}$$

(a) Show that the position vector when $t = \frac{\pi}{3} \, \text{s}$ is given by $\mathbf{r} = \frac{1}{6}\left(\sqrt{3}\pi\mathbf{i} + \pi\mathbf{j}\right)$. [5]

(b) Find the velocity vector of P and an expression for the speed of P at time t seconds in its simplest form. [4]

6 Types of force, resolving forces and forces in equilibrium

Essential facts and formulae

Facts

Types of force
- Weight – the force of gravity: $W = mg$
- Friction – the force that opposes motion
- Normal reaction – the force between a body and the surface on which it acts
- Tension – the resistive force in a string
- Thrust – the resistive force provided by a spring or rod

Modelling assumptions
- Strings – light and inextensible
- Objects/masses – assumed to be particles (i.e. point masses)
- Rods – uniform and light

Formulae

Resolution of forces
Replacing a single force with two forces acting at right angles

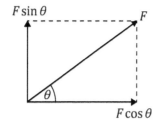

Force F can be replaced by two components at right angles to each other:
- A horizontal component $F \cos \theta$
- A vertical component $F \sin \theta$

Replacing two forces acting at right angles with a single force

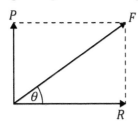

Using Pythagoras' theorem, so we obtain $F^2 = P^2 + R^2$

Hence $F = \sqrt{P^2 + R^2}$ and angle to the horizontal, $\theta = \tan^{-1}\left(\dfrac{P}{R}\right)$

Questions

1 Four horizontal forces of magnitude 6 N, 8 N, P N and R N acting at a point and in the same plane are in equilibrium. The direction of each force is shown in the diagram.

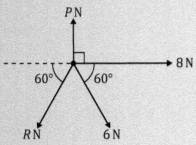

Find the exact magnitude of forces P and R. [4]

2 The diagram shows a sign attached to a point A. It is supported by two light rods AB and AC. The rod AC is horizontal and the rod AB is inclined at an angle of α to the horizontal, where $\sin \alpha = 0.6$.

The mass of the sign is 12 kg. Calculate:

(a) the thrust in the rod AB, [3]

(b) the tension in the rod AC. [3]

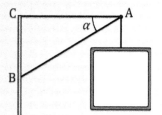

3 The diagram shows four coplanar forces acting at point O. The forces are all in equilibrium. Calculate the size of force Q and the size of angle θ giving your answer to two decimal places. [5]

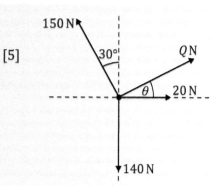

A lamp is hung from a point P and it is supported by two light rods PQ and PR. Rod PQ is horizontal and rod PR is inclined at an angle θ to the horizontal, where $\sin \theta = 0.5$. The mass of the lamp is 2 kg.

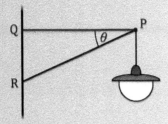

(a) Explain the difference between a thrust and a tension in a light rod. [2]

(b) Redraw the diagram marking all the forces acting. [2]

(c) Calculate the thrust in rod PR. [3]

(d) Calculate the tension in rod PQ. [3]

(e) Explain why the modelling assumption that the rods are light was made in this question. [1]

7 Forces and Newton's laws

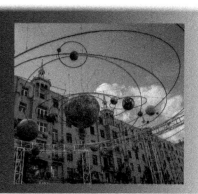

Essential facts and formulae

Facts

Newton's laws of motion

First law – a particle will remain at rest or will continue to move with constant speed in a straight line unless acted upon by some external force.

Second law – a resultant force produces an acceleration, according to the formula

Force = mass × acceleration.

Third law – every action has an equal and opposite reaction.

Finding a resultant force or acceleration

Resolve all the forces in two perpendicular directions and then use Pythagoras' theorem to find the resultant of these two forces at right angles. Use trigonometry to find the angle and clearly state the direction.

Use $a = \dfrac{F}{m}$ to find the acceleration which will be in the direction of the resultant force.

Formulae

Motion on an inclined plane

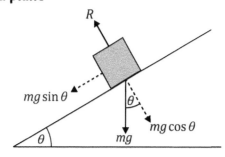

The component of the weight parallel to the slope = $mg \sin \theta$

The component of the weight at right angles to the slope = $mg \cos \theta$

Friction opposes motion

It can increase up to a certain maximum value called limiting friction or F_{MAX}.

Limiting friction, $F = \mu R$ which can also be written as $F_{\text{MAX}} = \mu R$, where μ is the coefficient of friction and R is the normal reaction between the surfaces.

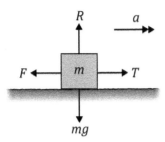

If $T > F$ there will be an unbalanced force and the mass will accelerate.

By Newton's 2nd law $\qquad\qquad ma = T - F$

Once moving, the mass will experience the maximum frictional force given by

$$F_{\text{MAX}} = \mu R$$

The forces are balanced in the vertical direction, so $R = mg$

Questions

1 A particle of mass 12 kg lies on a rough horizontal surface. The coefficient of friction between the particle and the surface is 0.8. The particle is at rest. It is then subjected to a horizontal tractive force of magnitude 75 N. Determine the magnitude of the frictional force acting on the particle, giving a reason for your answer. [5]

The diagram shows an object A, of mass 6 kg, lying on a rough horizontal table.

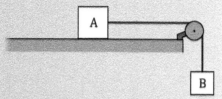

The object A is connected by means of a *light*, inextensible string passing over a smooth pulley at the edge of the table, to another object B, of mass 4 kg, hanging freely.

The coefficient of friction between object A and the table is 0.4. Initially, the system is held at rest with the string just taut. The system is then released.

(a) Find the magnitude of the acceleration of object A and the tension in the string. [9]

(b) What assumption did the word 'light' in italic above enable you to make in your solution? [1]

3 A rough plane is inclined at an angle α to the horizontal where $\sin \alpha = \frac{3}{5}$. A body of mass 80 kg lies on the plane. The coefficient of friction between the body and the plane is μ.

(a) Find the normal reaction of the plane on the body. [2]

(b) The body is on the point of slipping down the plane. Find the value of μ. [4]

(c) Calculate the magnitude of the force acting along a line of greatest slope that will move the body up the plane with an acceleration of 0.7 m s^{-2}. [4]

A particle of mass 2 kg moves in a horizontal plane under two forces whose magnitude and direction are shown in the diagram.

Find:

(a) The magnitude and direction of the resultant force giving your answer to two decimal places. [4]

(b) The magnitude of the acceleration. [2]

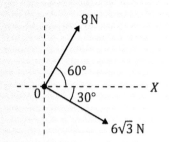

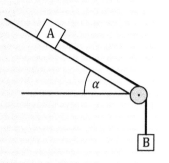 The diagram here shows two objects connected by means of a light inextensible string passing over a smooth light pulley. The pulley is fixed at the bottom of a rough plane inclined at an angle α to the horizontal, where $\tan \alpha = \frac{3}{4}$.

Object A, of mass 7 kg, lies on the inclined plane and object B, of mass 3 kg, is hanging freely. The coefficient of friction between the plane and object A is 0.6.

Initially, the objects are held at rest with the string just taut. The objects are then released so that A slides down the plane.

(a) Determine the magnitude of the frictional force acting on A. [3]

(b) Calculate the magnitude of the acceleration of the objects and the tension in the string. [7]

The diagram below shows a car of mass 1500 kg connected to a trailer of mass 600 kg by means of a rigid tow bar. The car is moving upwards along a slope inclined at an angle α to the horizontal, where $\sin \alpha = \frac{7}{25}$. A constant resistance of magnitude 400 N acts on the car and a constant resistance of 300 N acts on the trailer. The car's engine produces a constant forward force of 8400 N.

(a) Calculate the acceleration of the car, giving your answer correct to three decimal places. [5]

(b) Determine the tension in the tow bar. [4]

8 Projectile motion

Essential facts and formulae

Facts

Modelling assumptions:
- The body can be treated as a particle.
- No friction acts – the only force acting is gravity.
- The acceleration due to gravity, g, is constant.
- Motion is confined to the vertical plane.

Motion

The vertical component of the velocity is affected by gravity.

The horizontal component of the velocity is constant.

Formulae

The equations of motion (all of these must be memorised)

$$v = u + at \qquad\qquad v^2 = u^2 + 2as$$

$$s = ut + \tfrac{1}{2}at^2 \qquad\qquad s = \tfrac{1}{2}(u + v)t$$

> s = displacement
> u = initial velocity
> v = final velocity
> a = acceleration
> t = time

If the acceleration is zero (i.e. there is constant velocity), speed $= \dfrac{\text{distance}}{\text{time}}$

Motion of a particle projected at an angle α to the horizontal with velocity $U\,\mathrm{m\,s^{-1}}$

The horizontal component, $\qquad u_x = U \cos \alpha$

The vertical component, $\qquad u_y = U \sin \alpha$

$$\text{Time of flight} = \frac{2U \sin \alpha}{g}$$

$$\text{Range} = \frac{U^2 \sin 2\alpha}{g}$$

$$\text{Maximum height} = \frac{U^2 \sin^2 \alpha}{2g}$$

> Don't waste time trying to remember these formulae as you always need to derive them from first principles before using them.

Equation for the path of a projectile at any point (x, y) along its path is given by:

$$y = x \tan \alpha - \frac{gx^2(1 + \tan^2 \alpha)}{2U^2}$$

Questions

1 Points A and B lie on horizontal ground. At time $t = 0$ seconds, an object P is projected from A towards B such that AB is the range of P. The speed of projection is 24.5 m s^{-1} in a direction which is 30° above the horizontal.

(a) Calculate the range AB of the object P.   [5]

At time $t = 1$ second, another object Q is projected from B towards A with the same speed of projection 24.5 m s^{-1} and in a direction which is also 30° above the horizontal.

(b) Determine the height above the ground at which P and Q collide.   [5]

2 A golfer hits a ball from a point A with initial velocity of 35 m s^{-1} at an angle α above the horizontal where sin α = 0.8. The ball passes over a tree which is growing in front of a lake. The lake is 100 m wide, as shown in the diagram. The tree is at a horizontal distance of 17.5 m from A.

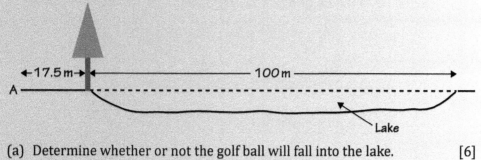

(a) Determine whether or not the golf ball will fall into the lake. [6]

(b) Find the magnitude and direction of the velocity of the ball as it passes over the tree. [7]

3 An arrow fired from a bow with an initial speed of 42 m s^{-1} hits a target 90 m away. Both the point of projection and the target are at the same horizontal height.

(a) Show that the horizontal range for a projectile is given by:

$$\frac{U^2 \sin 2\alpha}{g}$$

where U is the speed of projection and α is the angle of projection measured from the horizontal.   [5]

(b) Find the two possible values of angle α and state the angle that would produce the greatest time the arrow spends in the air.   [4]

(c) Describe one modelling assumption you have made in order to arrive at your answers to (a) and (b).   [1]

9 Differential equations

Essential facts and formulae

Facts

Formation of simple differential equations

If a quantity x has a rate of **increase** in x that is proportional to x, then this can be written as a differential equation by including a constant of proportionality k as

$$\frac{dx}{dt} = kx. \qquad k > 0$$

If a quantity x has a rate of **decrease** in x that is proportional to x, then this can be written as a differential equation by including a constant of proportionality k as

$$\frac{dx}{dt} = -kx. \qquad k > 0$$

Separating variables and integrating

If $\quad \dfrac{dm}{dt} = -km \quad$ and $\quad m = m_0$ at $t = 0$,

then variables can be separated and integrated as follows:

$$\int \frac{1}{m} \, dm = -k \int dt$$

$$\therefore \ln m = -kt + c \tag{1}$$

When $t = 0$, $m = m_0$

Substituting these values in (1), we obtain $\qquad c = \ln m_0$

Hence we obtain $\qquad\qquad\qquad\qquad\qquad \ln m = -kt + \ln m_0$

$$\ln m - \ln m_0 = -kt$$

$$\ln \left(\frac{m}{m_0} \right) = -kt$$

Taking exponentials of both sides

$$\frac{m}{m_0} = e^{-kt} \qquad\qquad \text{Hence} \qquad m = m_0 e^{-kt}$$

Similarly, if

$$\frac{dP}{dt} \propto f(P) \qquad\qquad \text{Then} \qquad \frac{dP}{dt} = kf(P)$$

$$\text{So} \quad \int \frac{1}{f(P)} \, dP = \int k \, dt$$

> This is done to remove the ln from the left-hand side.

> Usually, $f(P) = P^n$ where n is a constant.

> $k > 0$ if P increases with time and
> $k < 0$ if P decreases with time.

Questions

 An object of mass 0.5 kg is thrown vertically upwards with initial speed 24 m s^{-1}. The velocity of the object at time t seconds is v m s^{-1}. During the upward motion, the object experiences a resistance to motion R N, where R is proportional to v. When the velocity of the object is 0.2 m s^{-1} the resistance to motion is 0.08 N.

(a) Show that the upward motion of the object satisfies the differential equation

$$\frac{dv}{dt} = -9.8 - 0.8v$$ [3]

(b) Find an expression for t in terms of v. [6]

(c) Determine the value of t when the object is at the highest point of the motion. Give your value for t to 2 decimal places. [2]

 At time $t = 0$, a particle of mass 6 kg is projected vertically upwards from a point A with a speed of 24.5 m s^{-1}. The resistance acting on the particle has magnitude $3v$ N, where v m s^{-1} is the speed of the particle at time t s.

(a) (i) Show that v satisfies the equation $2\frac{dv}{dt} = -19.6 - v$. [2]

(ii) Find an expression for v in terms of t. [8]

(b) Determine the time when the particle reaches its maximum height. [2]

(c) Find an expression for x in terms of t, where x m is the distance of the particle from A at time t s. [4]

3 A horizontal force of 1400 N acts on a mass of 20 kg which causes the mass to move in a horizontal line. A resistive force acts on the mass which is proportional to the time t seconds. At time t seconds, the velocity of the mass is v m s^{-1} and at time $t = 4$ seconds, it is moving with velocity 24 m s^{-1} and acceleration -2 m s^{-2}.

(a) Show that v satisfies the differential equation:

$$\frac{dv}{dt} = 70 - 18t.$$ [4]

(b) Find an expression for v in terms of t. [6]

(c) Find the velocity of the mass after 5 seconds. [2]

4 A patient is injected with a drug which produces an initial concentration of 2 mg per litre of the drug in their bloodstream.

After a time t hours there are x mg per litre of drug in their bloodstream.

It is known that x decreases at a rate proportional to x. After a time of 1 hour, $x = 1.6$ mg per litre.

(a) Calculate the value of x after three hours. [6]

(b) Calculate the time to the nearest minute after which $x = 0.5$. [4]

Sample Test Paper Unit 3
Pure Mathematics B

(2 hour 30 minutes)

1 Prove by contradiction, for every real number x such that $0 \leq x \leq \frac{\pi}{2}$,
$$\sin x + \cos x \geq 1 .$$
[4]

2 Solve the equation:
$$|2x + 1| = 3|x - 2|$$
[4]

3 The sum of eight terms of an arithmetic progression is 1084. If the third term is 206, find the common difference and first term of the series. [5]

4 Use small angle approximation to find the small positive root of the equation:
$$12 \sin x - 2 \cos x = 8x^2$$
Give your answer to two decimal places. [3]

5 The function f has domain $(-\infty, 10]$ as is defined by:
$$f(x) = e^{5 - \frac{x}{2}} + 6$$
(a) Find an expression for $f^{-1}(x)$. [4]
(b) Write down the domain of f^{-1}. [2]

6 If $f(x) = \dfrac{11 + x - x^2}{(x+1)(x-2)^2}$, find $f'(x)$ when $x = 0$ [7]

7 Curve C has equation $3x^2y^3 = 216$.

(a) Find an expression for $\dfrac{dy}{dx}$. [3]

(b) Point P(a, b) lies on the curve and the gradient at this point is $-\dfrac{4}{9}$.

Find the values of a and b. [4]

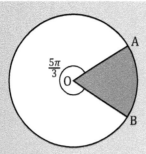

The diagram above shows a circle with centre O and points A and B on the circumference of the circle. The radius of the circle is 9 cm and the reflex angle AOB is $\frac{5\pi}{3}$ radians.

(a) Calculate the exact length of the minor arc AB.   [3]

(b) Determine the shaded area giving your answer to 2 decimal places.   [2]

⑨ Find $\int_0^1 \sqrt{4 - 4x^2}\, dx$ [6]

10 The population in a village was 2000 in 2010 and was 2350 in 2015.
If the rate of increase in the population x is proportional to x, calculate the
population in 2019. [7]

11 (a) Find the values of θ in the range $0° \leq x \leq 360°$ that satisfy the equation:

$$2\operatorname{cosec}^2\theta + \cot\theta = 8$$

Give your values of θ to two decimal places. [4]

(b) (i) Express $5\cos\theta + 12\sin\theta$ in the form $R\cos(\theta - \alpha)$, where R and α
are constants with $R > 0$ and $0° \leq \alpha \leq 90°$.

(ii) Use your results to part (i) to find the least value of

$$\frac{1}{5\cos\theta + 12\sin\theta + 25}$$

Write down a value of θ for which this least value occurs. [6]

12 Given that for $t > 0$,

$$x = \ln t, \quad y = 6t^2$$

(a) Find and simplify an expression for $\frac{dy}{dx}$ in terms of t. [4]

(b) Find the value of t for which $\frac{d^2y}{dx^2} = 6$. [4]

13 (a) On the same set of axes, sketch the graphs of $y = e^{-x}$ and $y = x$ for $0 \le x \le 1$. [3]

(b) (i) Hence, show that the equation $e^{-x} = x$ has only one root. [1]

(ii) Estimate the root to one decimal place. [1]

(iii) Using the value of your root as x_0 use the recurrence relation:

$$x_{n+1} = e^{-x_n}$$

to find the root to two decimal places. [6]

Sample Test Paper Unit 4
Applied Mathematics B

(1 hour 45 minutes)

Section A – Statistics

1 The probability that a person has a flu jab is 0.3. The probability that a person picked at random catches the flu after being given the flu jab is 0.1 and the probability of catching the flu when they have not been given the flu jab is 0.6.

    (a) Draw a tree diagram to represent these probabilities. Indicate the probabilities on your tree diagram. [3]

    (b) Use your tree diagram to find the probability that a person picked at random catches the flu. [3]

    (c) Given that a person has the flu, what is the probability that they were given the flu jab. [3]

2 A car approaches a level crossing and if the barrier just closes, the time in minutes to wait for barrier to be raised again follows a continuous uniform distribution over the interval [2, 6].

Find:

    (a) The expected wait in minutes for the barrier to be raised. [2]

    (b) The variance in minutes for the barrier to be raised. [2]

    (c) The probability that a person waits between 3 and 4 minutes for the barrier to be raised assuming it has just been lowered. [2]

Wild bird food is filled in sacks marked 12 kg by a machine. The weight of bird food in each sack can be considered to be normally distributed with a mean of 12.2 kg.

The machine filling the sacks ensures that 95% of the sacks have a weight of between 12.1 and 12.3 kg.

(a) Find the variance of the distribution of weights of bird food giving your answer to two significant figures. [6]

(b) The manager takes a sample of 15 sacks of the bird food filled by the machine and he measures the weight of each sack accurately and the following results are obtained:

12.2 12.4 12.3 12.2 12.2 12.1 12.5 12.0 12.2 12.5 12.3 12.5
12.0 12.1 12.3

Assuming that the standard deviation of the actual weights is 0.05, test at the 5% level of significance to decide whether the machine is filling the bags with a mean of 12.2 kg of bird food. [4]

(a) Suggest a value for the product moment correlation coefficient, r, for
each of the following distributions. [4]

(i)

(ii)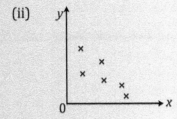

(iii)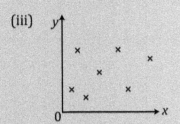

(iv)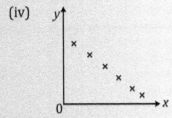

(b) A physicist takes readings of two quantities X and Y which she suspects
are correlated. She uses 40 pairs of values to calculate the product
moment correlation coefficient and finds that it is -0.55. Test at the
10% level of significance, stating your hypotheses clearly, to see if the
two quantities are correlated. [6]

Section B – Mechanics

 A particle moves in a straight line with a velocity v m s^{-1} at time t s where:

$$v = -6 \sin 3t$$

Calculate the distance travelled by the particle between $t = 0$ and $t = \dfrac{2\pi}{3}$ s. [5]

 A particle moves on a horizontal plane such that its velocity vector **v** m s^{-1} at time t s is given by:

$$\mathbf{v} = 2 \sin 2t\mathbf{i} - 12 \cos 3t\mathbf{j}$$

(a) Find the acceleration vector of the particle at time t s. [3]

(b) Given that, at $t = 0$, the particle has position vector $(2\mathbf{i} + 3\mathbf{j})$ m, find the position vector of the particle when $t = \dfrac{\pi}{2}$. [4]

3 A horizontal force of 10 N acts on a box of mass 5 kg on a rough horizontal surface.

This force causes the box to accelerate with an acceleration of 0.8 m s^{-2}.

(a) Draw a diagram showing all the forces acting on the box. [2]

(b) Calculate the magnitude of the normal reaction acting on the box. [1]

(c) Find the magnitude of the frictional force acting on the box. [2]

(d) Find the value of the coefficient of friction between the box and the surface giving your answer to one decimal place. [2]

4 A particle is projected with a speed U m s^{-1} at an angle of $\theta°$ to the horizontal. It then hits the horizontal ground a distance of R m from its point of projection.

(a) Show that the time the particle spends in the air, t s, is given by:

$$t = \frac{2U \sin \theta}{g}$$

(b) Find an expression for the horizontal distance travelled in terms of U, θ, and g. [4]

(c) Describe one modelling assumption you have made when obtaining your answer to part (b). [1]

(d) If the horizontal distance travelled is 32 m and the angle of projection to the horizontal was 60°, find:

(i) The speed of projection. [2]

(ii) The time to reach the maximum height. [2]

(iii) The maximum height reached. [2]

5 The rate of change of population of a small village is proportional to the square root of the population P. If the constant of proportionality is k, and at $t = 0$ years the population of the village is 1600:

(a) Find an expression for the population P in terms of k and t. [4]

(b) Find the population to the nearest integer after 4 years if $k = 0.25$. [2]

Answers

Unit 3 Pure Mathematics B
1 Proof

1 Assume that there is an integer for which a^2 is even and a is odd.
Now if a is odd there exists an integer n for which $a = 2n + 1$.
Squaring both sides, we obtain
$$a^2 = (2n + 1)^2 = 4n^2 + 4n + 1$$
So $a^2 = 2(n^2 + 2n) + 1$, which is always an odd number.
We therefore have the contradiction that a^2 is both odd and even.
Hence the original proposition, that if a^2 is even, then a is even, is true.

> For an integer value of n, the value of $2n + 1$ is always odd.

> $2(n^2 + 2n)$ is always even as it has 2 as a factor. Adding 1 to this will always result in an odd number.

2 Assume that there are positive integer solutions to the equation $x^2 - y^2 = 1$.
Now $x^2 - y^2 = 1$, so $(x + y)(x - y) = 1$
As x and y are positive integers then
$$[x + y = 1 \text{ and } x - y = 1] \quad \text{or} \quad [x + y = -1 \text{ and } x - y = -1].$$
Adding the first two equations, we obtain $2x = 2$ so $x = 1$ and if $x = 1, y = 0$.
Adding the second two equations, we obtain $2x = -2$ so $x = -1$ and $x = -1, y = 0$.
Now x and y are positive integers and here we have proved that this is not always the case. There is a contradiction, so the assumption is incorrect, so there are no positive integer solutions.

3 We assume that there is a real value of x for which $\sin x - \cos x < 1$.
Now $\sin x$ is positive and $\cos x$ is negative for $\frac{\pi}{2} \leq x \leq \pi$ as the following graph shows:
$$\sin x - \cos x < 1.$$
Squaring both sides, we obtain
$$(\sin x - \cos x)^2 < 1^2$$
$$\sin^2 x - 2\sin x \cos x + \cos^2 x < 1$$
However, $\sin^2 x + \cos^2 x = 1$
So, $-2\sin x \cos x + 1 < 1,$
hence $-2\sin x \cos x < 0.$
But as $\sin x$ is positive and $\cos x$ is negative, $-2\sin x \cos x$ cannot be less than 0 (i.e. negative) which is a contradiction.
Hence the original statement, that $\sin x - \cos x \geq 1$, is correct.

> Drawing a graph of $\sin x$ and $\cos x$ in the domain specified allows you to see the values $\sin x$ and $\cos x$ can take.

> Always clearly state what the contradiction is.

4 Assume that there is a value of θ for which
$$\sin \theta - \cos \theta > \sqrt{2}$$
Then squaring both sides, we have
$$(\sin \theta - \cos \theta)^2 > 2$$
$$\sin^2 \theta + 2\sin \theta \cos \theta + \cos^2 \theta > 2$$
Now $\sin^2 \theta + \cos^2 \theta = 1$ and $2\sin \theta \cos \theta = \sin 2\theta$
Hence $1 + \sin 2\theta > 2$, so $\sin 2\theta > 1$
This is a contradiction as the sine of any angle is ≤ 1.
Hence $\sin \theta + \cos \theta \leq \sqrt{2}$ is true.

Answers **Unit 3 Pure Mathematics B**

⑤ Assume that there is a real value of x such that $\left|x + \dfrac{1}{x}\right| < 2$.

Then squaring both sides, we have: $\left(x + \dfrac{1}{x}\right)^2 < 4$

$$x^2 + 2 + \dfrac{1}{x^2} < 4$$

Multiplying both sides by x^2, we obtain

$$x^4 + 2x^2 + 1 < 4x^2$$
$$x^4 - 2x^2 + 1 < 0$$
$$(x^2 - 1)(x^2 - 1) < 0$$

$(x^2 - 1)^2 < 0$, which is impossible since the square of a real number cannot be negative.

As this is a contradiction, the assumption is incorrect, so $\left|x + \dfrac{1}{x}\right| \geq 2$.

The modulus of x is written as $|x|$ and this means you take the numerical value of x ignoring the signs.

Notice there is an x^2. In equations such as this we can multiply both sides by x^2 to remove the fraction.

⑥ Assume that there is a positive and real value of x such that

$$4x + \dfrac{9}{x} < 12$$
$$4x^2 + 9 < 12x$$
$$4x^2 - 12x + 9 < 0$$
$$(2x - 3)^2 < 0$$

Now $(2x - 3)^2$ is always positive or zero for any real value of x, so $(2x - 3)^2$ cannot be < 0.

This is a contradiction, so the assumption is false and $4x + \dfrac{9}{x} \geq 12$.

Remove the denominator by multiplying both sides by x.

Factorise the quadratic inequality.

2 Algebra and functions

① (a) $\dfrac{x^2 - 4}{x^2 + x - 2} = \dfrac{(x + 2)(x - 2)}{(x + 2)(x - 1)} = \dfrac{x - 2}{x - 1}$

(b) $\dfrac{x^2 + 2x + 1}{3x^2 + 12x + 9} = \dfrac{(x + 1)(x + 1)}{3(x^2 + 4x + 3)} = \dfrac{(x + 1)(x + 1)}{3(x + 1)(x + 3)} = \dfrac{x + 1}{3(x + 3)}$

Make sure that you cancel any common factors in the numerator and denominator.

② $\dfrac{5x^2 + 6x + 7}{(x - 1)(x + 2)^2} \equiv \dfrac{A}{(x + 2)^2} + \dfrac{B}{x + 2} + \dfrac{C}{x - 1}$

$5x^2 + 6x + 7 \equiv A(x - 1) + B(x + 2)(x - 1) + C(x + 2)^2$

Let $x = 1$,   so $18 = 9C$,   giving $C = 2$
Let $x = -2$, so $15 = -3A$,   giving $A = -5$
Let $x = 0$,   so $7 = -A - 2B + 4C$,   giving $B = 3$

Hence $\dfrac{5x^2 + 6x + 7}{(x - 1)(x + 2)^2} \equiv \dfrac{-5}{(x + 2)^2} + \dfrac{3}{x + 2} + \dfrac{2}{x - 1}$

Notice the repeating linear factor in the denominator.

Alternatively, equate coefficients of x^2:
$5 = B + C$
$B = 5 - C = 3$

Checking by letting $x = 2$

$$\text{LHS} = \dfrac{5(2)^2 + 6(2) + 7}{(2 - 1)(2 + 2)^2} = \dfrac{39}{16}$$

$$\text{RHS} = \dfrac{-5}{(2 + 2)^2} + \dfrac{3}{2 + 2} + \dfrac{2}{2 - 1} = \dfrac{-5}{16} + \dfrac{3}{4} + 2 = \dfrac{39}{16}$$

Hence   LHS = RHS

Partial fractions are: $\dfrac{-5}{(x + 2)^2} + \dfrac{3}{x + 2} + \dfrac{2}{x - 1}$

$A = -5$, $B = 3$ and $C = 2$

3 $\dfrac{3x}{(1+x)^2(2+x)} \equiv \dfrac{A}{(1+x)^2} + \dfrac{B}{1+x} + \dfrac{C}{2+x}$

$$3x \equiv A(2+x) + B(1+x)(2+x) + C(1+x)^2$$

Let $x = -2$ so $C = -6$
Let $x = -1$ so $A = -3$
Equating coefficients of x^2, $0 = B + C$, $0 = B - 6$, $B = 6$

Hence $f(x) = -\dfrac{3}{(1+x)^2} + \dfrac{6}{1+x} - \dfrac{6}{2+x}$

4 $f(x+4)$ represents a translation by $\begin{pmatrix} -4 \\ 0 \end{pmatrix}$, the $\frac{2}{3}$ represents a stretch parallel to the y-axis and the negative sign represents a reflection in the x-axis.

So the coordinates of the points of intersection with the x-axis become $(-5, 0)$ and $(3, 0)$ and the stationary point changes to $(-1, 4)$.

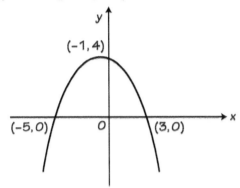

5 (a) (i) Domain of f is larger than the domain of g so domain of fg is restricted.
Hence $D(fg) = \left(0, \dfrac{\pi}{4}\right)$

> Notice the domain of g (i.e. $\left(0, \frac{\pi}{4}\right)$) lies inside the domain of f. Hence, we need to restrict the domain of the composite function to the domain of g.

(ii) $fg(x) = \ln(\tan x)$
As $D(fg) = \left(0, \dfrac{\pi}{4}\right)$

when $x \to 0$, $\ln(\tan x) \to \ln(\tan 0) \to -\infty$

when $x \to \dfrac{\pi}{4}$, $\ln(\tan x) = \ln\left(\tan \dfrac{\pi}{4}\right) = \ln 1 = 0$

Hence $R(fg) = (-\infty, 0]$

> Notice the need for the square bracket for 0 as the range can equal zero.

(b) (i) $\ln(\tan x) = -0.4$
Taking exponentials of both sides, we obtain
$$\tan x = e^{-0.4}$$
$$= 0.67032 \ldots$$
$$x = 0.59$$

> Remember to set your calculator to radians when working out x.

(ii) In order for the equation $fg(x) = k$ to have a solution, k must be in the range of fg. Hence, we can choose any value for k that is outside the range $(-\infty, 0]$
So we could choose $k = 1$.

6 (a) t is the number of days after which the person was infected, so it cannot be negative.
(b) When $t = 0$,
$$f(0) = \frac{6000}{1 + 5999e^{-0.8 \times 0}}$$

> Remember $e^0 = 1$

$$f(0) = \frac{6000}{1 + 5999e^0} = \frac{6000}{6000} = 1$$

This is the first person who was infected.

(c) $f(5) = \dfrac{6000}{1 + 5999e^{-0.8 \times 5}}$

$= 54$ (to the nearest whole number)

(d) 40% of 6000 = 2400

$$2400 = \frac{6000}{1 + 5999e^{-0.8t}}$$

$$1 + 5999e^{-0.8t} = \frac{6000}{2400}$$

$$1 + 5999e^{-0.8t} = 2.5$$

$$5999e^{-0.8t} = 1.5$$

$$e^{-0.8t} = \frac{1.5}{5999}$$

$$e^{-0.8t} = 2.5004 \ldots \times 10^{-4}$$

Taking ln of both sides $-0.8t = -8.293 \ldots$

$$t = 10.4 \text{ days (1 d.p.)}$$

> Remember $\ln(e^{-x}) = -x$

7 $|2x + 3| = x$

$2x + 3 = \pm x$

$2x + 3 = x$ or $2x + 3 = -x$

$x = -3$ or $x = -1$

> The alternative method could also be used where you square both sides to remove the modulus sign and then solve the resulting quadratic equation.

8   (a) To find the points of intersection with the x-axis, put $f(x) = 0$.

$0 = x^2 - 6x + 8$

$0 = (x - 4)(x - 2)$

$x = 2$ or 4 (note that 4 is not in the domain so we ignore this point).

To find the minimum point we need to complete the square.

$x^2 - 6x + 8 = (x - 3)^2 - 9 + 8$

$= (x - 3)^2 - 1$

Hence minimum point occurs at $(3, -1)$

To find the point of intersection with the y-axis, substitute $x = 0$ into the equation of the curve. When $x = 0$, $y = 8$.

> Watch out here. It would be easy to draw the whole curve, but you should only draw the part covered by the domain, i.e.$(-\infty, 3)$.

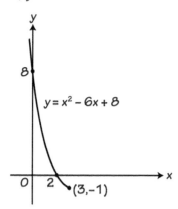

(b) $R(f) = (-1, \infty)$

(c) $y = (x - 3)^2 - 1$

$y + 1 = (x - 3)^2$

$\pm\sqrt{y + 1} = x - 3$

$x = \pm\sqrt{y + 1} + 3$

The domain of $f^{-1}(x)$ is the same as the range of $f(x)$.

Domain of $f^{-1}(x)$ is $(-1, \infty)$

9 (a) The graph of $y = \cos x$ is shown below:

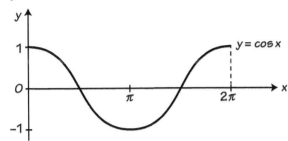

The negative section of the graph is reflected in the x-axis to produce the graph $y = |\cos x|$.

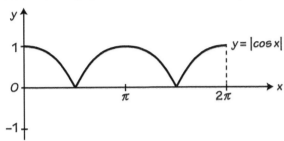

(b) To produce the graph of $y = f(x) + 1$ you translate the graph $y = |\cos x|$ by $\binom{0}{1}$.
Hence, we obtain

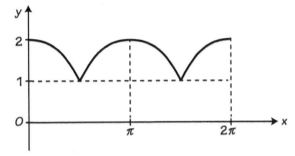

10 (a) $|5x - 8| = 2$

$5x - 8 = \pm 2$

$5x - 8 = 2$ or $5x - 8 = -2$

$5x = 10$ or $5x = 6$

$x = 2$ or $x = 1.2$

(b) (i)

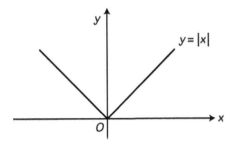

> You could have used the alternative method where you square both sides of the equation to obtain a quadratic equation which is then factorised and solved to find the two values of x.

(ii) $y = -x^2 + 7x - 10$

To find the coordinates of the points of intersection with the x-axis, we let $y = 0$.

$0 = -x^2 + 7x - 10$

$x^2 - 7x + 10 = 0$

$(x - 5)(x - 2) = 0$

$x = 5$ or $x = 2$.

> It is easier to factorise this if you rearrange the equation so that x^2 is positive.

When $x = 0$, $y = -0^2 + 7(0) - 10 = -10$
The curve for the modulus will cut the y-axis at $(0, 10)$

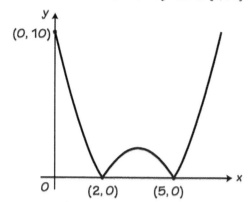

11   (a)   Let $y = \ln(3x - 2) + 5$

$y - 5 = \ln(3x - 2)$

$e^{y-5} = 3x - 2$

$e^{y-5} + 2 = 3x$

$x = \dfrac{e^{y-5} + 2}{3}$

$f^{-1}(x) = \dfrac{e^{x-5} + 2}{3}$

> Rearrange the equation to make x the subject of the equation. Then replace y with x and replace the subject of the equation by $f^{-1}(x)$.

> Convert from the logarithmic to the index form of this relationship.

(b)   The domain of f^{-1} is the same as the range of f.

$R(f) = [5, \infty)$

Hence $D(f^{-1}) = [5, \infty)$

12   (a) $gf(x) = 4[f(x)]^3 + 7 \quad = 4(e^x)^3 + 7$

$\qquad\qquad\qquad\qquad\quad = 4e^{3x} + 7$

> The domain gf is the set of all x in the domain of f for which $f(x)$ is in the domain of g.
> f has domain $[0, \infty)$ and range $[1, \infty)$, and g has domain $(-\infty, \infty)$. The domain of gf is the domain of f.

(b) $D(gf) = [0, \infty)$

When $x = 0$, $gf(0) = 4e^0 + 7 = 11$

As $x \to \infty$, $gf(x) \to \infty$

Hence $R(gf) = [11, \infty)$

> Substitute values from the domain into $gf(x)$ to find its maximum and minimum values.

(c)   (i) $gf(x) = 18$

$4e^{3x} + 7 = 18$

$4e^{3x} = 11$

$e^{3x} = \dfrac{11}{4}$

$3x = \ln\dfrac{11}{4}$

$3x = 1.0116$

$x = 0.3372$

$x = 0.337$ (correct to three decimal places)

(ii) $R(gf) = [11, \infty)$, these are the maximum and minimum y-values if a graph of the composite function were drawn.

There is no y-value for the function outside this range.

$k = 8$ would be one of the many values that would yield no solution.

> As the range is $[11, \infty)$, the line $y = k$ would not have any solutions for values of k outside the range.
> Any values of k less than 11 would suffice.

3 Sequences and series

1. (a) See section on 'Proof of the formula for the sum of an arithmetic series' on page 60 of the A2 Pure book.

   (b) $a = 4$ and $d = 2$

   $S_n = 460$

   $$S_n = \frac{n}{2}\left[2a + (n-1)d\right]$$

   $$460 = \frac{n}{2}\left[8 + (n-1)2\right]$$

   > Substituting $a = 4$, $d = 2$ and $S_n = 460$ into the formula for S_n.

   $$920 = n(2n + 6)$$
   $$920 = 2n^2 + 6n$$

   > Divide both sides by 2.

   $$460 = n^2 + 3n$$
   $$n^2 + 3n - 460 = 0$$
   $$(n + 23)(n - 20) = 0$$

   $n = 20$ as the other value would mean a negative number of terms.

   > This is quite a hard quadratic to factorise. The two factors need to be close together to give $+3n$ in the middle.

   (c)
   $$t_5 = a + 4d$$
   $$9 = a + 4d \qquad (1)$$
   $$t_6 = a + 5d$$
   $$t_{10} = a + 9d$$
   $$t_6 + t_{10} = a + 5d + a + 9d = 2a + 14d$$
   $$2a + 14d = 42$$
   $$21 = a + 7d \qquad (2)$$

   Solving equations (1) and (2) simultaneously
   equation (2) − equation (1) gives $3d = 12$
   $$d = 4$$
   Substituting this value of d into equation (1)
   $$9 = a + 4 \times 4$$
   $$9 = a + 16$$
   $$a = -7$$

   Hence first term $a = -7$ and common difference $d = 4$

2. (a) $t_{n+1} = 2t_n + 1$
   $$t_4 = 2t_3 + 1$$
   $$63 = 2t_3 + 1$$
   $$t_3 = 31$$
   $$t_3 = 2t_2 + 1$$
   $$31 = 2t_2 + 1$$
   $$t_2 = 15$$
   $$t_2 = 2t_1 + 1$$
   $$15 = 2t_1 + 1$$
   $$t_1 = 7$$

   (b) 6 043 582 is even but all the terms of the sequence are odd.

   2 × (an even or odd number) always results in an even number and adding a one to an even number will always make an odd number.

3 The terms of the series are $1 + 4 + 7 + 10 + \ldots$

First term $a = 1$ and common difference, $d = 3$.

$$S_n = \frac{n}{2}\left[2a + (n-1)d\right]$$

$$S_{100} = \frac{100}{2}\left[2 \times 1 + (100 - 1)3\right]$$

$$S_{100} = 50[2 + 99 \times 3]$$

$$S_{100} = 14\,950$$

The series is obtained by substituting 1 then 2 then 3 into the expression.

This formula is obtained from the formula booklet.

Remember to do the multiplication before the addition in the square bracket.

4 (a) (i) $\dfrac{1}{\sqrt{1 + 2x}} = (1 + 2x)^{-\frac{1}{2}}$

$$(1 + x)^n = 1 + nx + \frac{n(n-1)x^2}{2!} + \frac{n(n-1)(n-2)\,x^3}{3!} + \ldots$$

$$(1 + 2x)^{-\frac{1}{2}} = 1 + \left(-\frac{1}{2}\right)2x + \frac{\left(-\frac{1}{2}\right)\left(-\frac{3}{2}\right)(2x)^2}{2} + \ldots$$

$$= 1 - x + \frac{3x^2}{2}$$

(ii) Expansion is valid for $|2x| < 1$ so $|x| < \dfrac{1}{2}$.

(b)
$$4 + 15x - x^2 = 6\left(1 - x + \frac{3x^2}{2}\right)$$
$$4 + 15x - x^2 = 6 - 6x + 9x^2$$
$$10x^2 - 21x + 2 = 0$$
$$(10x - 1)(x - 2) = 0$$

Hence $x = 0.1$ (note the other solution, $x = 2$, is outside the range of x for which the expansion is valid).

5 (a) $(1 + x)^n = 1 + nx + \dfrac{n(n-1)x^2}{2!} + \dfrac{n(n-1)(n-2)\,x^3}{3!} + \ldots$

$$(1 - x)^{-\frac{1}{2}} = 1 + \left(-\frac{1}{2}\right)(-x) + \frac{\left(-\frac{1}{2}\right)\left(-\frac{3}{2}\right)(-x)^2}{2} + \ldots$$

$$= 1 + \frac{x}{2} + \frac{3x^2}{8} + \ldots$$

This expansion is valid for $|x| < 1$.

(b) When $x = \dfrac{1}{10}$,
$$\left(\frac{9}{10}\right)^{\frac{1}{2}} \approx 1 + \frac{1}{20} + \frac{3}{800}$$

$$\approx \frac{843}{800}$$

Now $\sqrt{10} = 10^{\frac{1}{2}}$ and $\left(\dfrac{9}{10}\right)^{-\frac{1}{2}} = \left(\dfrac{10}{9}\right)^{\frac{1}{2}} = \dfrac{\sqrt{10}}{3}$

Hence $\dfrac{843}{800} = \dfrac{\sqrt{10}}{3}$

So, $\sqrt{10} = 3 \times \dfrac{843}{800}$

$$= \frac{2529}{800}$$

4 Trigonometry

❶ $\dfrac{\sin(A-B)}{\cos A \cos B} = \dfrac{\sin A \cos B - \cos A \sin B}{\cos A \cos B}$

$= \dfrac{\sin A \cos B}{\cos A \cos B} - \dfrac{\cos A \sin B}{\cos A \cos B}$

$= \dfrac{\sin A}{\cos A} - \dfrac{\sin B}{\cos B}$

$= \tan A - \tan B$

> The expansion of $\sin(A-B)$ is obtained from the formula booklet.

> $\dfrac{\sin x}{\cos x} = \tan x$

❷ (a) $3\cos\theta + 4\sin\theta \equiv R\cos(\theta - \alpha)$

$3\cos\theta + 4\sin\theta \equiv R\cos\theta\cos\alpha + R\sin\theta\sin\alpha$

Hence $R\cos\alpha = 3$ and $R\sin\alpha = 4$

$\tan\alpha = \dfrac{4}{3}$ so $\alpha = 53.1°$

$R = \sqrt{3^2 + 4^2} = \sqrt{25} = 5$

Hence $3\cos\theta + 4\sin\theta = 5\cos(\theta - 53.1°)$

> $\dfrac{R\sin\alpha}{R\cos\alpha} = \tan\alpha = \dfrac{4}{3}$

> Only the positive value of $\sqrt{25}$ is used because $R > 0$.

(b) $\dfrac{1}{3\cos\theta + 4\sin\theta + 7} = \dfrac{1}{5\cos(\theta - 53.1°) + 7}$

The greatest value of $\cos(\theta - 53.1°)$ is 1.

So least value of $\dfrac{1}{3\cos\theta + 4\sin\theta + 7}$ is $\dfrac{1}{5 + 7} = \dfrac{1}{12}$

This occurs when

$\cos(\theta - 53.1°) = 1$

So $\theta - 53.1° = 0$

or $\theta = 53.1°$

> Other values of θ are possible.

❸ If $\phi = 360° - \theta$ then $\cos\theta = \cos\phi$ (e.g. if $\theta = 60°$ then $\phi = 300°$, then $\cos 60°$ and $\cos 300°$ both equal 0.5).

However, $\sin 60° = \dfrac{\sqrt{3}}{2}$ and $\sin 300° = -\dfrac{\sqrt{3}}{2}$ so $\sin\theta \neq \sin\phi$

Hence the statement is false.

❹ (a)

(b) Let $f(x) = \cos^{-1} x - 5x + 1$

$f(0.4) = \cos^{-1} 0.4 - 5(0.4) + 1 = 0.16$

$f(0.5) = \cos^{-1} 0.5 - 5(0.5) + 1 = -0.45$

As there is a change of sign, one of the roots must lie between 0.4 and 0.5.

5 $\cos(x - \alpha) - \cos \alpha = \cos x \cos \alpha + \sin x \sin \alpha - \cos \alpha$

Now, for small angles $\cos x = 1 - \dfrac{x^2}{2}$

Hence, $\cos(x - \alpha) - \cos \alpha = \left(1 - \dfrac{x^2}{2}\right)\cos \alpha + \sin x \sin \alpha - \cos \alpha$

$$= x \sin \alpha - \dfrac{x^2}{2}\cos \alpha$$

> Notice that the question refers to a small angle which means the small angle formula needs to be used.

> As x is a small angle, $\sin x$ is approximately equal to x.

6 (a) Let $\sin x + \cos x = R \cos(x - \alpha)$
$$= R \cos x \cos \alpha + R \sin x \sin \alpha$$

$R \cos \alpha = 1$ and $R \sin \alpha = 1$ so $\dfrac{R \sin \alpha}{R \cos \alpha} = 1$ and $\tan \alpha = 1$ giving $\alpha = 45°$

Hence, we can write $\quad \sin x + \cos x + 3 = \sqrt{2}\cos(x - 45°) + 3$
The maximum value of $\sqrt{2}\cos(x - 45°) + 3 = \sqrt{2} + 3 = 3 + \sqrt{2}$
The minimum value of $\sqrt{2}\cos(x - 45°) + 3 = -\sqrt{2} + 3 = 3 - \sqrt{2}$

(b) Maximum value of $\dfrac{1}{f(x)}$ occurs when $f(x)$ has its minimum value.

Hence maximum value is $\dfrac{1}{3 - \sqrt{2}}$

Minimum value of $\dfrac{1}{f(x)}$ occurs when $f(x)$ has its maximum value.

Hence minimum value is $\dfrac{1}{3 + \sqrt{3}}$

7 $\qquad\qquad 2\sin(x - 60°) = \cos(\theta + 60°)$
$2(\sin x \cos 60° - \cos x \sin 60°) = \cos x \cos 60° - \sin x \sin 60°$

Now $\cos 60° = \dfrac{1}{2}$ and $\sin 60° = \dfrac{\sqrt{3}}{2}$

Hence, $\quad 2\left(\dfrac{1}{2}\sin x - \dfrac{\sqrt{3}}{2}\cos x\right) = \dfrac{1}{2}\cos x - \dfrac{\sqrt{3}}{2}\sin x$

$\sin x - \sqrt{3}\cos x = \dfrac{1}{2}\cos x - \dfrac{\sqrt{3}}{2}\sin x$

$2\sin x - 2\sqrt{3}\cos x = \cos x - \sqrt{3}\sin x$

$2\sin x + \sqrt{3}\sin x = \cos x + 2\sqrt{3}\cos x$

$\sin x \left(2 + \sqrt{3}\right) = \cos x \left(1 + 2\sqrt{3}\right)$

$\tan x = \dfrac{1 + 2\sqrt{3}}{2 + \sqrt{3}}$

$= \dfrac{(1 + 2\sqrt{3})(2 - \sqrt{3})}{(2 + \sqrt{3})(2 - \sqrt{3})}$

$= \dfrac{2 - \sqrt{3} + 4\sqrt{3} - 6}{4 - 3}$

$= 3\sqrt{3} - 4$

> Look in the formula booklet for both these expansions.

> Note that we need to use exact values here.

> Multiply through by 2 to remove the denominators.

5 Differentiation

1 $\dfrac{dV}{dt} = 0.05$

$$V = \frac{4}{3}\pi r^3 \text{ so } \frac{dV}{dr} = 4\pi r^2$$

$$A = 4\pi r^2 \text{ so } \frac{dA}{dr} = 8\pi r$$

$$\frac{dA}{dt} = \frac{dA}{dr} \times \frac{dV}{dt} \times \frac{dr}{dV}$$

$$= 8\pi r \times 0.05 \times \frac{1}{4\pi r^2}$$

When $r = 8$, $\quad \dfrac{dA}{dt} = 64\pi \times 0.05 \times \dfrac{1}{4\pi(8)^2}$

$$= 0.0125 \text{ m}^2 \text{ per minute}$$

2 (a) Let $y = (3x^2 - 2x)^7$

$$\frac{dy}{dx} = 7(6x - 2)(3x^2 - 2x)^6$$

> The quick way to differentiate this is to first multiply by the power 7, then multiply by the derivative of the bracket and then multiply this by the original contents with the bracket raised to one less than the original power.

(b) Let $y = \sqrt{3x^3 - 4}$

$$y = (3x^3 - 4)^{\frac{1}{2}}$$

$$\frac{dy}{dx} = \frac{1}{2}(9x^2)(3x^3 - 4)^{-\frac{1}{2}}$$

$$\frac{dy}{dx} = \frac{9x^2}{2\sqrt{3x^3 - 4}}$$

> Again, the quick method is used here but use the method you find most comfortable.

(c) Let $y = x^3 \ln 3x$

Using the Product rule: If $y = f(x)g(x)$,

$$\frac{dy}{dx} = f(x)g'(x) + g(x)f'(x)$$

$$= x^3 \frac{3}{3x} + \ln 3x(3x^2)$$

$$= x^2 + 3x^2 \ln x$$

> The derivative of $\ln 3x$ is $\frac{3}{3x}$ which can be simplified to $\frac{1}{x}$.

(d) Let $y = \dfrac{e^{2x}}{(2x + 1)^5}$

Using the Quotient rule, if $y = \dfrac{f(x)}{g(x)}$,

$$\frac{dy}{dx} = \frac{f'(x)g(x) - f(x)g'(x)}{(g(x))^2}$$

$$= \frac{2e^{2x}(2x + 1)^5 - e^{2x}(2)(5)(2x + 1)^4}{(2x + 1)^{10}}$$

$$= \frac{2e^{2x}(2x + 1)^5 - 10e^{2x}(2x + 1)^4}{(2x + 1)^{10}}$$

$$= \frac{2e^{2x}(2x + 1)^4(2x + 1 - 5)}{(2x + 1)^{10}}$$

$$= \frac{2e^{2x}(2x - 4)}{(2x + 1)^6}$$

$$= \frac{4e^{2x}(x - 2)}{(2x + 1)^6}$$

> The Quotient rule is looked up in the formula booklet.

> Try to take as many factors out as you can as the question asks you to simplify your answer.

3 (a) Let $y = \ln(9x^3 - 2x + 1)$

$$\frac{dy}{dx} = \frac{27x^2 - 2}{9x^2 - 2x + 1}$$

(b) $\frac{d}{dx}\left(e^{f(x)}\right) = e^{f(x)} f'(x)$

> You need to write $\sqrt{x}$ in index form (i.e. $x^{\frac{1}{2}}$) before differentiating.

So, $\frac{d}{dx}\left(e^{\sqrt{x}}\right) = e^{\sqrt{x}}\left(\frac{1}{2}x^{-\frac{1}{2}}\right)$

$$= \frac{1}{2\sqrt{x}} e^{\sqrt{x}}$$

(c) Let $y = \dfrac{a - b\cos x}{a + b\sin x}$

> This is a quotient, so the Quotient rule must the used. The Quotient rule is included in the formula booklet.

$$\frac{dy}{dx} = \frac{b\sin x(a + b\sin x) - (a - b\cos x)(b\cos x)}{(a + b\sin x)^2}$$

$$= \frac{ab\sin x + b^2\sin^2 x - ab\cos x + b^2\cos^2 x}{(a + b\sin x)^2}$$

4 $\dfrac{6 + x - 9x^2}{x^2(x + 2)} \equiv \dfrac{A}{x^2} + \dfrac{B}{x} + \dfrac{C}{x + 2}$

$6 + x - 9x^2 \equiv A(x + 2) + Bx(x + 2) + Cx^2$

> You could have alternatively equated coefficients to find the values of A, B and C.

Let $x = -2$, $6 - 2 - 36 = 4C$,   hence $C = -8$

Let $x = 0$, $6 = 2A$,   hence $A = 3$

Let $x = 1$, $-2 = 3A + 3B + C$,   hence $B = -1$

Hence $f(x) = \dfrac{3}{x^2} - \dfrac{1}{x} - \dfrac{8}{x + 2}$

Writing this in index form, we have $f(x) = 3x^{-2} - x^{-1} - 8(x + 2)^{-1}$

Hence $f'(x) = -6x^{-3} + x^{-2} + 8(x + 2)^{-2}$

$$f'(x) = \frac{-6}{x^3} + \frac{1}{x^2} + \frac{8}{(x + 2)^2}$$

When $x = 2$, $f'(x) = \dfrac{-6}{2^3} + \dfrac{1}{2^2} + \dfrac{8}{(2 + 2)^2}$

$$= -\frac{3}{4} + \frac{1}{4} + \frac{1}{2} = 0$$

The gradient is 0 at $x = 2$ so this is a stationary point.

5 $x^3 - 2xy^2 + y^3 = 5$

> As this equation involves terms in x and y, we need to differentiate implicitly to find the gradient.

Differentiating implicitly with respect to x we obtain:

$$3x^2 - 2x(2y)\frac{dy}{dx} - 2y^2 + 3y^2\frac{dy}{dx} = 0$$

> Need to rearrange this to find $\frac{dy}{dx}$ on its own.

$$\frac{dy}{dx}\left(3y^2 - 4xy\right) = 2y^2 - 3x^2$$

$$\frac{dy}{dx} = \frac{2y^2 - 3x^2}{3y^2 - 4xy}$$

When $x = 2$ and $y = 1$, $\dfrac{dy}{dx} = \dfrac{2 - 12}{3 - 8} = 2$

> The coordinates and the gradient are substituted into the equation for a straight line $y - y_1 = m(x - x_1)$.

Equation of tangent at (2, 1) is $y - 1 = 2(x - 2)$

$$y = 2x - 3$$

6 $\frac{dx}{dt} = 2$ and $\frac{dy}{dt} = 15t^2$

> Note $\frac{dt}{dx} = \frac{1}{\frac{dx}{dt}}$ so you must remember to invert $\frac{dx}{dt}$.

Hence, $\frac{dy}{dx} = \frac{dy}{dt} \times \frac{dt}{dx} = 15t^2 \times \frac{1}{2} = \frac{15}{2}t^2$

As point P lies on C and has parameter p.
Equation of the tangent to C at P is given by:
$$y - y_1 = m(x - x_1)$$
$$y - 5p^3 = \frac{15}{2}p^2(x - 2p)$$
$$2y - 10p^3 = 15p^2x - 30p^3$$
$$2y = 15p^2x - 20p^3$$

7 (a) Let $y = \ln(\cos x)$ and $u = \cos x$

Hence $y = \ln u$ so $\frac{dy}{du} = \frac{1}{u} = \frac{1}{\cos x}$

$\frac{du}{dx} = -\sin x$

$\frac{dy}{dx} = \frac{dy}{du} \times \frac{du}{dx} = \frac{1}{\cos x} \times (-\sin x) = -\tan x$

(b) Let $y = \tan^{-1}\left(\frac{x}{2}\right)$

$$\frac{d}{dx}\left(\tan^{-1}(f(x))\right) = \frac{1}{1+(f(x))^2} \times f'(x)$$

$$\frac{dy}{dx} = \frac{1}{1+\left(\frac{x}{2}\right)^2} \times \frac{1}{2} = \frac{1}{1+\frac{x^2}{4}} \times \frac{1}{2} = \frac{1}{\frac{4+x^2}{4}} \times \frac{1}{2} = \frac{2}{4+x^2}$$

(c) Let $y = e^{4x}(2x - 1)^4$
Using the Product rule
$$\frac{dy}{dx} = e^{4x}\left(4 \times 2(2x-1)^3\right) + (2x-1)^4 4e^{4x}$$
$$= 8e^{4x}(2x-1)^3 + 4e^{4x}(2x-1)^4$$
$$= 4e^{4x}(2x-1)^3(2 + 2x - 1)$$
$$= 4e^{4x}(2x-1)^3(2x+1)$$

8 (a) $\frac{d}{dx}\left(\tan^{-1}(f(x))\right) = \frac{1}{1+(f(x))^2} \times f'(x)$

$$\frac{d(\tan^{-1}(5x))}{dx} = \frac{1}{1+(5x)^2} \times 5 = \frac{5}{1+25x^2}$$

(b) Let $y = e^{x^2}$
$$\frac{d}{dx}\left(e^{f(x)}\right) = e^{f(x)}f'(x)$$
So, $\frac{d}{dx}\left(e^{x^2}\right) = e^{x^2}(2x) = 2xe^{x^2}$

(c) $\frac{d(x^3\ln x)}{dx} = x^3\left(\frac{1}{x}\right) + \ln x(3x^2) = x^2 + 3x^2\ln x$

(d) $\frac{d\left(\frac{5-x^2}{3x^2-1}\right)}{dx} = \frac{(3x^2-1)(-2x)-(5-x^2)(6x)}{(3x^2-1)^2}$

$$= \frac{-6x^3+2x-30x+6x^3}{(3x^2-1)^2} = \frac{-28x}{(3x^2-1)^2}$$

9 $x = 2t - \sin 2t$

$$\frac{dx}{dt} = 2 - 2\cos 2t$$

$y = \cos 3t$

$$\frac{dy}{dt} = -3\sin 3t$$

$$\frac{dy}{dx} = \frac{dy}{dt} \times \frac{dt}{dx}$$

So

$$\frac{dy}{dx} = \frac{(-3\sin 3t)}{(2 - 2\cos 2t)}$$

When $t = \frac{\pi}{2}$

$$\frac{dy}{dx} = \frac{\left(-3\sin\frac{3\pi}{2}\right)}{\left(2 - 2\cos\frac{2\pi}{2}\right)}$$

$$= \frac{3}{4}$$

> First differentiate the parametric equations by means of the Chain rule.

> $\sin\frac{3\pi}{2} = -1$ and $\cos\pi = -1$

10 (a) $f(x) = x^2 e^x$

$$f'(x) = x^2 e^x + e^x(2x)$$
$$f'(x) = xe^x(x + 2)$$

(b) $f'(1) = 1e^1(1 + 2) = 3e$

> This is a product, so the Product rule is used.

6 Coordinate geometry in the (x, y) plane

1 $x^2 + 3xy + 3y^2 + 13$

Differentiating with respect to x we obtain

$$2x + (3x)\frac{dy}{dx} + y(3) + 6y\frac{dy}{dx} = 0$$

giving

$$\frac{dy}{dx} = \frac{-2x - 3y}{3x + 6y}$$

At the point (2, 1),

$$\frac{dy}{dx} = \frac{-2(2) - 3(1)}{3(2) + 6(1)} = -\frac{7}{12}$$

Gradient of the normal $= \frac{12}{7}$

Equation of the normal at the point (2, 1) is

$$y - 1 = \frac{12}{7}(x - 2)$$
$$7y - 7 = 12x - 24$$
$$7y - 12x + 17 = 0$$

> Using the fact that the product of the gradients of the tangent and normal is −1.

> Use $y - y_1 = m(x - x_1)$ to find the equation of the normal.

2 (a) $\frac{dx}{dt} = 2t$ and $\frac{dy}{dt} = 3t^2$

Hence

$$\frac{dy}{dx} = \frac{dy}{dt} \times \frac{dt}{dx} = 3t^2 \times \frac{1}{2t} = \frac{3t}{2}$$

At $P(p^2, p^3)$, $\frac{dy}{dx} = \frac{3p}{2}$

Equation of tangent at P is

$$y - p^3 = \frac{3p}{2}(x - p^2)$$
$$2y - 2p^3 = 3px - 3p^3$$
$$3px - 2y = p^3$$

(b) $3px - 2y = p^3$, so when $p = 2$, $6x - 2y = 8$

As Q lies on the tangent its coordinates will satisfy the equation of the tangent.

So substituting $x = q^2$, $y = q^3$ into the equation gives $6q^2 - 2q^3 = 8$

Hence $3q^2 - q^3 = 4$ so $q^3 - 3q^2 + 4 = 0$

Let $f(q) = q^3 - 3q^2 + 4$

$\quad\quad f(1) = 1^3 - 3(1)^2 + 4 = 2$

$\quad\quad f(-1) = (-1)^3 - 3(-1)^2 + 4 = 0$ hence $(q + 1)$ is a factor

Let $(q + 1)(aq^2 + bq + c) \equiv q^3 - 3q^2 + 4$

Equating coefficients of q^3, we obtain $a = 1$

Equating coefficients independent of q, we obtain $c = 4$

Equating coefficients of q^2 gives $b + a = -3$, so $b = -4$

Hence $q^3 - 3q^2 + 4 = (q + 1)(q^2 - 4q + 4) = (q + 1)(q - 2)(q - 2)$

Now $(q + 1)(q - 2)(q - 2) = 0$

Solving we obtain $q = -1, 2$

The value $q = 2$ relates to point P, so $q = -1$

> Note this is an identity, i.e. true for all values of q.

3 (a) $\dfrac{dx}{dt} = 8 \cos 2t$ and $\dfrac{dy}{dt} = -6 \sin 2t$

Hence $\dfrac{dy}{dx} = \dfrac{dy}{dt} \times \dfrac{dt}{dx} = (-6 \sin 2t) \times \dfrac{1}{8 \cos 2t} = -\dfrac{3}{4} \tan 2t$

(b) (i) $\dfrac{d^2y}{dx^2} = -\dfrac{3}{2} \sec^2 2t$

(ii) Now $\sec 2t = \dfrac{1}{\cos 2t}$

Hence, $\dfrac{d^2y}{dx^2} = -\dfrac{3}{2} \times \dfrac{1}{\cos^2 2t}$

Since $\cos 2t = \dfrac{y}{3}$ we can write $\dfrac{d^2y}{dx^2} = -\dfrac{3}{2} \times \dfrac{1}{\cos^2 2t} = -\dfrac{3}{2} \times \dfrac{9}{y^2} = -\dfrac{27}{2y^2}$

4 (a) $x^4 + x^2y + y^2 = 13$

Differentiating implicitly with respect to x, we obtain

$$4x^3 + x^2(1)\dfrac{dy}{dx} + y(2x) + 2y\dfrac{dy}{dx} = 0$$

$$\dfrac{dy}{dx}(x^2 + 2y) = -4x^3 - 2xy$$

$$\dfrac{dy}{dx} = \dfrac{-4x^3 - 2xy}{x^2 + 2y}$$

At the point $(-1, 3)$, $\dfrac{dy}{dx} = \dfrac{-4x^3 - 2xy}{x^2 + 2y} = \dfrac{-4 + 6}{1 + 6} = \dfrac{2}{7}$

(b) Now $x = p^2$ and $y = 2p$, hence $\dfrac{dy}{dp} = 2$ and $\dfrac{dx}{dp} = 2p$

Hence $\dfrac{dy}{dx} = \dfrac{dy}{dp} \times \dfrac{dp}{dx} = 2 \times \dfrac{1}{2p} = \dfrac{1}{p}$

> Note that this is the gradient of the tangent to the curve.

Gradient of normal $= -p$

Equation of the normal is $y - 2p = -p(x - p^2)$

Hence, $\quad\quad\quad y + px = 2p + p^3$

When $y = 0$, $x = b$ so $\quad b = 2 + p^2$

Since $p^2 > 0$, $b > 2$

5 $\frac{dx}{dt} = 2t$ and $\frac{dy}{dt} = 12t^3 + 24t^2$

Hence $\frac{dy}{dx} = \frac{dy}{dt} \times \frac{dt}{dx} = (12t^3 + 24t^2) \times \frac{1}{2t} = 6t^2 + 12t$

When $t = -1$, $\frac{dy}{dx} = 6(-1)^2 + 12(-1) = -6$

When $t = -1$, $x = (-1)^2 - 4 = -3$ and $y = 3(-1)^4 - 8(-1)^3 = -5$

Hence equation of tangent is:
$$y - y_1 = m(x - x_1)$$
$$y + 5 = -6(x + 3)$$
$$y + 5 = -6x - 18$$
$$y = -6x - 23 \text{ or } y + 6x + 23 = 0$$

7 Integration

1 Let $u = e^x$ so $\frac{du}{dx} = e^x$ and $du = e^x dx$

Changing the limits, we have, when $x = 1$, $u = e$ and when $x = 0$, $u = 1$.

$$\int_0^1 \frac{e^x}{1 + e^x} dx = \int_1^e \frac{1}{1 + u} du$$
$$= [\ln(1 + u)]_1^e$$
$$= [\ln(1 + e) - \ln 2]$$
$$= \ln\frac{1 + e}{2}$$

2 (a) $\int e^{\frac{x}{2}} dx = \frac{e^{\frac{x}{2}}}{\frac{1}{2}} + c = 2e^{\frac{x}{2}} + c$

(b) $\int \frac{5x}{5x^2 + 9} dx = \frac{1}{2}\int \frac{10x}{5x^2 + 9} dx$

$$= \frac{1}{2}\ln(5x^2 + 9) + c$$

> You need to spot that the derivative of the denominator would give $10x$ and that it is possible to turn the numerator into $10x$. When the derivative of the denominator is the same as the numerator, the integral is the ln of the denominator.

3 Let $u = 2x + 1$, so $\frac{du}{dx} = 2$ and $dx = \frac{du}{2}$

Now $x = \frac{u - 1}{2}$

> When there is a bracket raised to a power, we usually let u be equal to the contents of the bracket.

Substituting these parts into the integral, we obtain:

$$\int \frac{u - 1}{2}\left(\frac{1}{u^3}\right)\frac{du}{2} = \frac{1}{4}\int \frac{u - 1}{u^3} du$$

$$= \frac{1}{4}\int (u^{-2} - u^{-3}) du$$

$$= \frac{1}{4}\left(\frac{u^{-1}}{-1} - \frac{u^{-2}}{-2}\right) + c$$

$$= \frac{1}{4}\left(-\frac{1}{u} + \frac{1}{2u^2}\right) + c$$

$$= \frac{1}{4}\left(\frac{1}{2u^2} - \frac{1}{u}\right) + c$$

We now substitute $u = 2x + 1$ back into the above.

Hence, we have, integral $= \frac{1}{4}\left(\frac{1}{2(2x + 1)^2} - \frac{1}{(2x + 1)}\right) + c$

4 Let $u = 4x + 1$, so $\dfrac{du}{dx} = 4$ and $dx = \dfrac{du}{4}$

Now $4x = u - 1$, so $2x = \dfrac{u - 1}{2}$

When $x = \dfrac{1}{4}$, $u = 4\left(\dfrac{1}{4}\right) + 1 = 2$ and when $x = -\dfrac{1}{4}$, $u = 4\left(-\dfrac{1}{4}\right) + 1 = 0$

The integral now becomes $\displaystyle\int_0^2 \dfrac{u - 1}{2} \times u^6 \dfrac{du}{4} = \dfrac{1}{8}\int_0^2 (u^7 - u^6)\, du$

$$= \dfrac{1}{8}\left[\dfrac{u^8}{8} - \dfrac{u^7}{7}\right]_0^2$$

$$= \dfrac{1}{8}\left[\left(\dfrac{2^8}{8} - \dfrac{2^7}{7}\right) - (0 - 0)\right]$$

$$= \dfrac{12}{7} = 1\dfrac{5}{7}$$

5 $\displaystyle\int_0^2 \dfrac{6}{(2x - 3)^2}\, dx = 6\int_0^2 \dfrac{1}{(2x - 3)^2}\, dx$

$$= 6\int_0^2 (2x - 3)^{-2}\, dx$$

$$= 6\left[\dfrac{(2x - 3)^{-1}}{(-1)(2)}\right]_0^2$$

$$= -3\left[\dfrac{1}{2x - 3}\right]_0^2$$

$$= -3\left[1 - \left(-\dfrac{1}{3}\right)\right]$$

$$= -4$$

6 (a) Notice the curves are not labelled.
The curve with equation $y = \dfrac{1}{2}\sin 2x$ cuts the x-axis at $x = k$.
Equation of x-axis is $y = 0$.

Hence $0 = \dfrac{1}{2}\sin 2x$.

When $x = 0$ $\sin 2x = 0$ but this is the point where the curve cuts the origin, so we ignore this value of x.
Now $\sin \pi = 0$ so $2x = \pi$, hence $x = \dfrac{\pi}{2}$.
Hence $k = \dfrac{\pi}{2}$.

(b) Shaded area = area under curve $y = \sin x$ between 0 and $\dfrac{\pi}{2}$, minus the area under the curve $y = \dfrac{1}{2}\sin 2x$ between 0 and $\dfrac{\pi}{2}$.
Rather than write them as separate integrals we can combine them like this to save time:

$$\text{Shaded area} = \int_0^{\frac{\pi}{2}} \left(\sin x - \dfrac{1}{2}\sin 2x\right) dx$$

$$= \left[-\cos x + \dfrac{1}{4}\cos 2x\right]_0^{\frac{\pi}{2}}$$

$$= \left[\left(-\cos\dfrac{\pi}{2} + \dfrac{1}{4}\cos 2\dfrac{\pi}{2}\right) - \left(-\cos 0 + \dfrac{1}{4}\cos 0\right)\right]$$

$$= \dfrac{1}{2}$$

7 $\dfrac{4x+1}{(x+1)^2(x-2)} = \dfrac{A}{(x+1)^2} + \dfrac{B}{x+1} + \dfrac{C}{x-2}$

$4x + 1 = A(x - 2) + B(x + 1)(x - 2) + C(x + 1)^2$

Let $x = 2$, $9 = 9C$ giving $C = 1$

Let $x = -1$, $-3 = -3A$ giving $A = 1$

Let $x = 0$, $1 = -2 - 2B + 1$ giving $B = -1$

The integral becomes:

$$\int_3^4 \left(\dfrac{1}{(x+1)^2} - \dfrac{1}{x+1} + \dfrac{1}{x-2} \right) dx$$

$$= \int_3^4 \left((x+1)^{-2} - \dfrac{1}{x+1} + \dfrac{1}{x-2} \right) dx$$

$$= \left[\dfrac{(x+1)^{-1}}{-1} - \ln(x+1) + \ln(x-2) \right]_3^4$$

$$= \left[-\dfrac{1}{x+1} - \ln(x+1) + \ln(x-2) \right]_3^4$$

$$= \left[\left(-\dfrac{1}{5} - \ln 5 + \ln 2 \right) - \left(-\dfrac{1}{4} - \ln 4 + \ln 1 \right) \right]$$

$$= \dfrac{1}{20} + \ln \dfrac{8}{5}$$

> The fraction needs to be converted to partial fractions before the expression can be integrated.

> Notice that both $\dfrac{1}{x+1}$ and $\dfrac{1}{x-2}$ have the differential of the denominator as the numerator.
> Hence the integral of each of these is the ln of the denominator.

8 $\displaystyle\int_1^2 \dfrac{6}{(2x-3)^3} dx = 6\int_1^2 \dfrac{1}{(2x-3)^3} dx$

$$= 6\int_1^2 (2x-3)^{-3} dx$$

$$= 6\left[\dfrac{(2x-3)^{-2}}{(-2)(2)} \right]_1^2$$

$$= -\dfrac{3}{2}\left[(2x-3)^{-2} \right]_1^2$$

$$= -\dfrac{3}{2}\left[1^{-2} - (-1)^{-2} \right]$$

$$= -\dfrac{3}{2}\left(1 - 1 \right)$$

$$= 0$$

9 (a) $\dfrac{3x}{(1+x)^2(2+x)} = \dfrac{A}{(1+x)^2} + \dfrac{B}{1+x} + \dfrac{C}{2+x}$

$3x = A(2 + x) + B(1 + x)(2 + x) + C(1 + x)^2$

Let $x = -2$,   so $C = -6$

Let $x = -1$,   so $A = -3$

Equating coefficients of x^2, $0 = B + C$, $0 = B - 6$, $B = 6$

Hence $f(x) = -\dfrac{3}{(1+x)^2} + \dfrac{6}{1+x} - \dfrac{6}{2+x}$

(b) $\displaystyle\int_0^1 f(x)\, dx = \int_0^1 \left(-\dfrac{3}{(1+x)^2} + \dfrac{6}{1+x} - \dfrac{6}{2+x} \right) dx$

$$= \int_0^1 \left(-3(1+x)^{-2} + \dfrac{6}{1+x} - \dfrac{6}{2+x} \right) dx$$

$$= \left[3(1+x)^{-1} + 6\ln(1+x) - 6\ln(2+x) \right]_0^1$$

$$= \left[\frac{3}{1+x} + 6\ln(1+x) - 6\ln(2+x) \right]_0^1$$

$$= \left[\left(\frac{3}{2} + 6\ln 2 - 6\ln 3 \right) - (3 + 6\ln 1 - 6\ln 2) \right]$$

$$= 0.226 \text{ (correct to 3 d.p.)}$$

10 Let $u = 1 + x^2$, so $\frac{du}{dx} = 2x$ and $dx = \frac{du}{2x}$

Hence $\int x\sqrt{1 + x^2}\, dx = \int xu^{\frac{1}{2}} \frac{du}{2x}$

$$= \int \frac{u^{\frac{1}{2}}}{2}\, du$$

$$= \frac{u^{\frac{3}{2}}}{3} + c$$

$$= \frac{(1 + x^2)^{\frac{3}{2}}}{3} + c$$

$$= \frac{1}{3}\sqrt{(1 + x^2)^3} + c$$

11 Let $u = 1 + x^4$, so $\frac{du}{dx} = 4x^3$ and $dx = \frac{du}{4x^3}$

$$\int \frac{x^3}{1 + x^4}\, dx = \int \frac{x^3}{u} \times \frac{du}{4x^3}$$

$$= \frac{1}{4} \int \frac{1}{u}\, du$$

$$= \frac{1}{4} \ln u + c$$

$$= \frac{1}{4} \ln(1 + x^4) + c$$

12 Let $u = \sin x$, $\frac{du}{dx} = \cos x$ and $dx = \frac{du}{\cos x}$

$$\int \frac{\cos x}{\sin^3 x}\, dx = \int \frac{\cos x}{u^3} \times \frac{du}{\cos x}$$

$$= \int u^{-3}\, du$$

$$= \frac{u^{-2}}{-2} + c$$

$$= -\frac{1}{2u^2} + c$$

$$= -\frac{1}{2\sin^2 x} + c$$

13 $\int \frac{5}{2 + 3x}\, dx = 5 \int \frac{1}{2 + 3x}\, dx$

$$= \frac{5}{3} \int \frac{3}{2 + 3x}\, dx$$

$$= \frac{5}{3} \ln(2 + 3x) + c$$

(14) (a) (i) $\int e^{-3x+5} \, dx = -\dfrac{e^{-3x+5}}{3} + c$

(ii) $\int x^2 \ln x \, dx$

Let $u = \ln x$ and $\dfrac{dv}{dx} = x^2$

The formula for integration by parts is used here. The formula is included in the formula booklet.

So, $\dfrac{du}{dx} = \dfrac{1}{x}$ and $v = \dfrac{x^3}{3}$

Hence, $\int x^2 \ln x \, dx = \dfrac{x^3}{3} \ln x - \int \dfrac{x^3}{3} \times \dfrac{1}{x} \, dx$

$= \dfrac{x^3}{3} \ln x - \dfrac{x^3}{9} + c$

(b) $\displaystyle\int_0^{\frac{1}{2}} \dfrac{x^2}{\sqrt{1-x^2}} \, dx$

Let $x = \sin \theta$, so $\dfrac{dx}{d\theta} = \cos \theta$ and $dx = \cos \theta \, d\theta$

Changing the limits: When $x = 0$, $\theta = 0$ and when $x = \dfrac{1}{2}$, $\theta = \dfrac{\pi}{6}$.

Hence, we have $\displaystyle\int_0^{\frac{1}{2}} \dfrac{x^2}{\sqrt{1-x^2}} \, dx = \int_0^{\frac{\pi}{6}} \dfrac{\sin^2 \theta \cos \theta}{\sqrt{1 - \sin^2 \theta}} \, d\theta$

Note that $1 - \sin^2 \theta = \cos^2 \theta$

$= \displaystyle\int_0^{\frac{\pi}{6}} \dfrac{\sin^2 \theta \cos \theta}{\cos \theta} \, d\theta$

$= \displaystyle\int_0^{\frac{\pi}{6}} \sin^2 \theta \, d\theta$

$= \displaystyle\int_0^{\frac{\pi}{6}} \dfrac{1 - \cos 2\theta}{2} \, d\theta$

$= \left[\dfrac{\theta}{2} - \dfrac{\sin 2\theta}{4} \right]_0^{\frac{\pi}{6}}$

$= \left(\dfrac{\pi}{12} - \dfrac{\sin \frac{\pi}{3}}{4} \right) - (0)$

$= \dfrac{\pi}{12} - \dfrac{\sqrt{3}}{8}$

(15) (a) $\int \dfrac{1}{1-x} \, dx = -\int \dfrac{-1}{1-x} \, dx$

$= -\ln|1 - x| + c$

(b) $\int (2x - 3)^5 \, dx = \dfrac{(2x - 3)^6}{2 \times 6} + c$

$= \dfrac{(2x - 3)^6}{12} + c$

(c) $\int 5 \sin (2x - 1) \, dx = 5 \int \sin (2x - 1) \, dx$

$= -\dfrac{5}{2} \cos (2x - 1) + c$

8 Numerical methods

① $\displaystyle\int_a^b y\,dx = \int_4^6 \frac{1}{3-\sqrt{x}}\,dx$

$$h = \frac{b-a}{n} = \frac{6-4}{4} = 0.5$$

> It is important to note that n is the number of strips and not the number of ordinates. Here there are 5 ordinates, so there will be 4 strips. Hence $n = 4$.

> This means that you start at the value of a (4 in this case) and go up in steps of h (0.5 here) until the value of b is reached (6 in this case).

When $x = 4$, $\quad y_0 = \dfrac{1}{3-\sqrt{4}} = 1$

$x = 4.5$, $\quad y_1 = \dfrac{1}{3-\sqrt{4.5}} = 1.138071187$

> Always work to at least one decimal place beyond that you are asked to find the answer to.

$x = 5$, $\quad y_2 = \dfrac{1}{3-\sqrt{5}} = 1.309016994$

$x = 5.5$, $\quad y_3 = \dfrac{1}{3-\sqrt{5.5}} = 1.527202251$

$x = 6$, $\quad y_4 = \dfrac{1}{3-\sqrt{6}} = 1.816496581$

$\displaystyle\int_a^b y\,dx \approx \frac{1}{2}h\{(y_0 + y_n) + 2(y_1 + y_2 + \ldots + y_{n-1})\}$

> $n = 4$

$\qquad \approx \dfrac{1}{2}(0.5)\{(1 + 1.816496581) + 2(1.138071187 + 1.309016994 + 1.527202251)\}$

$\qquad \approx 2.691269361$

$\qquad \approx 2.691$ (correct to three decimal places)

A useful check of your working which tests the reasonableness of your answer is to calculate the middle y value × range of integration.
In this case the middle y value is $y_2 = 1.309016994$ and the range of integration is $6 - 4 = 2$
The integral is approximately $2 \times 1.309016994 \approx 2.618$, which compares well with 2.691.

> **Grade boost**
> Remember to give your answer to the required number of decimal places or significant figures specified in the question.

② $\quad h = \dfrac{b-a}{n} = \dfrac{2-1}{4} = 0.25$

When $x = 1$, $\quad y_0 = \ln 1 = 0$

When $x = 1.25$, $\quad y_1 = \ln 1.25 = 0.223143551$

When $x = 1.50$, $\quad y_2 = \ln 1.50 = 0.405465108$

When $x = 1.75$, $\quad y_3 = \ln 1.75 = 0.559615788$

When $x = 2$, $\quad y_4 = \ln 2 = 0.693147181$

$\displaystyle\int_a^b y\,dx \approx \frac{1}{2}h\{(y_0 + y_n) + 2(y_1 + y_2 + \ldots + y_{n-1})\}$, where $h = \dfrac{b-a}{n}$

$\qquad \approx 0.5 \times 0.25\{(0 + 0.693147181) + 2(0.223143551 + 0.405465108 + 0.559615788)\}$

$\qquad \approx 0.3836995094$

$\qquad \approx 0.384$ (correct to three d.p.)

Since $\quad \displaystyle\int_1^2 \ln(x^2)\,dx = \int_1^2 2\ln x\,dx$

> Property of logs:
> $\ln(x^n) = n\ln x$

$\qquad\qquad = 2 \times 0.3836995094 = 0.767$ (correct to three d.p.)

Checking answer to $\displaystyle\int_1^2 \ln x\,dx$

Approximate value = middle y value × range of integration

$\qquad\qquad = y_2 \times (2-1)$

$\qquad\qquad = 0.405465108 \times 1 \approx 0.405$ (correct to three d.p.)

This compares with 0.386.

3 **(a)**

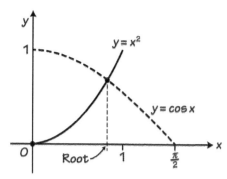

(b) Let $f(x) = \cos x - x^2$

$$f(0.8) = \cos 0.8 - 0.8^2 = 0.0567 \ldots$$
$$f(0.9) = \cos 0.9 - 0.9^2 = -0.1883 \ldots$$

There is a sign change, so the root lies between and including 0.8 and 0.9.

(c) $f(x) = \cos x - x^2$

$$f'(x) = -\sin x - 2x$$

$$x_{n+1} = x_n - \frac{f(x_n)}{f'(x_n)} = x_n - \frac{\cos x_n - x_n^2}{\sin x_n - 2x_n}$$

Start with $x_0 = 0.8$

$$x_1 = x_0 - \frac{\cos x_0 - x_0^2}{\sin x_0 - 2x_0} = 0.8 - \frac{\cos 0.8 - 0.8^2}{-\sin 0.8 - 2 \times 0.8}$$

> Note that the angles in the trig functions in this equation will be in radians. Remember to change your calculator to radians when working out the values.

This is quite a complicated equation, so it makes sense to use a calculator. It is best to think carefully about the bracketing you are going to use when you enter the calculation into the calculator.

For x_1 we can use the following bracketing:

$$x_1 = 0.8 - \left(\frac{(\cos 0.8 - 0.8^2)}{(-\sin 0.8 - 2 \times 0.8)} \right)$$

Remember to change the calculator to radians.
Enter into the calculator:

$$0.8 =$$

Then enter:

$$\text{Ans} - ((\cos(\text{Ans}) - \text{Ans}^2) \div (-\sin(\text{Ans}) - 2 \times \text{Ans})) =$$

Check that you get the following answer for x_1

$$x_1 = 0.824470434$$

Now press = to get the next value, x_2

$$x_2 = 0.8241323766$$

Now press = to get the next value, x_3

$$x_3 = 0.8241323123$$

Looking at the values for x_1, x_2 and x_3 you can see that the value to 3 decimal places is staying constant.

Hence root = 0.824 (correct to 3 d.p.)

4 (a)

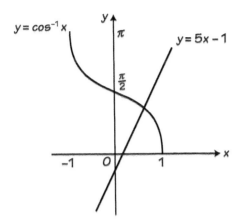

(b) $x_{n+1} = \frac{1}{5}\left(1 + \cos^{-1} x_n\right)$

$x_0 = 0.4$

Enter into the calculator:
$$0.4 =$$

Then enter:
$$1 + 5 \times (1 + \cos^{-1}(\text{Ans})) =$$

Check that you get the following answer for x_1
$$x_1 = 0.4318558961$$

Now press = to get the next value, x_2
$$x_2 = 0.4248493791$$

Now press = to get the next value, x_3
$$x_3 = 0.4264001662$$

Now press = to get the next value, x_4
$$x_4 = 0.4260574128$$

Hence $\qquad x_4 = 0.4261$ (correct to 4 d.p.)

Let $f(x) = \cos^{-1} x - 5x + 1$

Checking the sign of $f(x)$ for $x = 0.42605$ and $x = 0.42615$, we obtain:

$$f(0.42605) = \cos^{-1}(0.42605) - 5(0.42605) + 1 = 4.24 \times 10^{-4}$$

$$f(0.42615) = \cos^{-1}(0.42615) - 5(0.42615) + 1 = -1.86 \times 10^{-4}$$

As there is a change of sign between these two values,
$$\alpha = 0.4261 \text{ (correct to 4 d.p.)}$$

Unit 4 Applied Mathematics B
Section A: Statistics

1 Probability

1 (a) The following tree diagram is drawn.

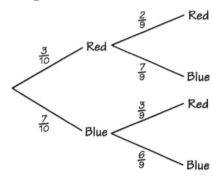

> Taking two counters together is considered the same as taking one counter and, without replacing it, taking another counter.

$$P(\text{Red and Red}) = \frac{3}{10} \times \frac{2}{9} = \frac{1}{15}$$

(b) $P(\text{Red and Blue})$ $= P(RB) + P(BR)$

$$= \frac{3}{10} \times \frac{7}{9} + \frac{7}{10} \times \frac{3}{9}$$

$$= \frac{42}{90}$$

$$= \frac{7}{15}$$

> Note that red and blue does not specify an order. There are two paths which need to be considered on the tree diagram.

> Remember to fully cancel fractions. Use your calculator to help you.

2 (a) $P(A \cap B) = P(A) \times P(B) = 0.4 \times 0.3 = 0.12$

(b) $P(A \cup B) = P(A) + P(B) - P(A \cap B) = 0.4 + 0.3 - 0.12 = 0.58$

(c) Probability that neither A nor B occurs $= P(A \cup B)' = 1 - P(A \cup B)$
$$= 1 - 0.58 = 0.42$$

> This could also be worked out in the following way:
> $P(A' \cap B') = (1 - 0.4)(1 - 0.3)$
> $= 0.42,$
> since A' and B' are independent.

(d)

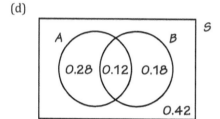

> We are asked for the probability of event A given that event $A \cup B$ has taken place. Hence we need to divide the probability of A by the probability of $A \cup B$. A Venn diagram can help present the probabilities already found.

$$P(A|A \cup B) = \frac{P(A)}{P(A \cup B)} = \frac{0.4}{0.58} = 0.69 \text{ (correct to two decimal places)}$$

3 (a) If events A and B were independent, then $P(A \cap B) = 0.3 \times 0.4 = 0.12$
Now $P(A \cap B) = P(A) + P(B) - P(A \cup B)$
$$= 0.3 + 0.4 - 0.5$$
$$= 0.2$$
A and B are not independent because $0.12 \neq 0.2$

(b)

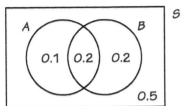

$$P(A|B') = \frac{P(A \cap B')}{P(B')}$$

$$= \frac{0.1}{0.1 + 0.5}$$

$$= \frac{0.1}{0.6}$$

$$= \frac{1}{6}$$

4 (a) (i) When events are mutually exclusive $P(A \cup B) = P(A) + P(B)$

$$= 0.5 + 0.3$$

$$= 0.8$$

(ii) For independent events, $P(A \cap B) = P(A) \times P(B)$

$$= 0.5 \times 0.3$$

$$= 0.15$$

$$P(A \cup B) = P(A) + P(B) - P(A \cap B)$$

$$= 0.5 + 0.3 - 0.15$$

$$= 0.65$$

(b) $P(A \cup B) = P(A) + P(B) - P(A \cap B)$

$$0.7 = 0.5 + 0.3 - P(A \cap B)$$

$$P(A \cap B) = 0.1$$

$$P(A \cap B) = P(A) P(B|A)$$

Hence, $P(B|A) = \dfrac{P(A \cap B)}{P(A)}$

$$= \frac{0.1}{0.5}$$

$$= 0.2$$

5 (a) $P(C') = \dfrac{48}{100} = \dfrac{12}{25}$

(b) $P(D' \cap C') = \dfrac{16}{100} = \dfrac{4}{25}$

(c)

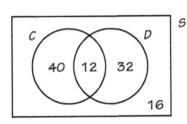

> This part of the question is easier to answer if a Venn diagram is drawn using the information in the table.

$$P(D \cup C) = \frac{84}{100} = \frac{21}{25}$$

6 (a)

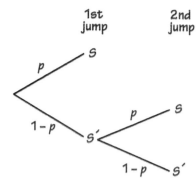

1st jump   2nd jump

Let S = event Marie is successful at the jump

$$P(\text{Clears } 1.7\,\text{m}) = p + (1 - p)p$$
$$= 2p - p^2$$

Now $P(\text{Clears } 1.7\,\text{m}) = 0.64$

So, $0.64 = 2p - p^2$

$$p^2 - 2p + 0.64 = 0$$
$$(p - 1.6)(p - 0.4) = 0$$
$$p = 1.6 \text{ or } 0.4$$

Probability cannot be greater than 1, so $p = 1.6$ is rejected.

Hence $p = 0.4$

(b) We know the first athlete is male so we are only considering how many field athletes there are from the male athletes.

Hence, we have $P(\text{First athlete is a male field athlete}) = \dfrac{9}{13 + 9}$

$$= \dfrac{9}{22}$$

The second athlete can be male or female, but must be a field athlete.

Now as a male field athlete has already been chosen, there are only 32 athletes to choose from and only 8 male field athletes.

Hence the probability of the second athlete being a field athlete, given that the first athlete was a male field athlete is:

$$\dfrac{(8 + 4)}{(13 + 7 + 8 + 4)} = \dfrac{12}{32}$$

Hence,

$P(\text{Given the first athlete is male, both are field athletes}) = \dfrac{9}{22} \times \dfrac{12}{32}$

$$= \dfrac{27}{176}$$

$$= 0.1534 \dots$$

7 (a) If A and B were independent events, the probability of winning both bids would be

$$P(A) \times P(B) = 0.6 \times 0.5 = 0.3$$

However, the probability of winning both is 0.2, so the events (i.e. winning bids) are dependent events. So this is a question about conditional probability.

$$P(A \cap B) = P(A)\,P(B|A)$$
$$0.2 = 0.5 \times P(B|A)$$
$$P(B|A) = 0.4$$

This formula is used when you are dealing with dependent events.

We can draw the Venn diagram, remembering to subtract the probability of the intersection of A and B (i.e. 0.2) from the probabilities of A and B respectively.

The total probability has to add up to 1

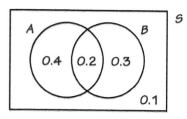

So $\qquad\qquad 1 - (0.4 + 0.2 + 0.3) = 0.1$

Probability of not winning either bid = 0.1

> Note the probability of not winning either bid is represented by the region $(A \cup B)'$ on the Venn diagram.

(b) P(exactly 1 bid) $= P(A) + P(B) - 2\,P(A \cap B)$

$\qquad\qquad\qquad = 0.6 + 0.5 - 2 \times 0.2$

$\qquad\qquad\qquad = 0.7$

(c) $P(B|A') = \dfrac{P(B \cap A')}{P(A')}$

$\qquad\quad = \dfrac{0.3}{(0.3 + 0.1)}$

$\qquad\quad = 0.75$

> Here you have to adapt the formula
> $$P(B|A) = \frac{P(A \cap B)}{P(B)}$$
> by changing A to B and B to A' to give
> $$P(B|A') = \frac{P(B \cap A')}{P(A')}$$

2 Statistical distributions

1 Required probability is P($W < 1000$).

$$P(W < 1000) = \frac{d - c}{b - a}$$

$$= \frac{1000 - 980}{1030 - 980}$$

$$= 0.4$$

2 (a) (i) A continuous uniform distribution with parameters [0, 12].

(ii) If $X \sim U(a, b)$, then

$$\text{Mean, E}(X) = \tfrac{1}{2}\big(a + b\big) = \tfrac{1}{2}\big(0 + 12\big) = 6$$

$$\text{Variance, Var}(X) = \tfrac{1}{12}\big(b - a\big)^2 = \tfrac{1}{12}\big(12 - 0\big)^2 = 12$$

(iii) Just because they are scheduled to depart at 12 min intervals does not mean they will, as they could be delayed.

(b) (i) This question requires careful thought. There are two scenarios to consider:

The first scenario is he waits between 9 and 12 minutes and gets on the first train as he is not distracted by his smartphone.

Probability of waiting between 9 and 12 minutes:

$$P(9 < X < 12) = \frac{d - c}{b - a} = \frac{12 - 9}{12 - 0} = \frac{1}{4}$$

Probability of **not** being distracted by smartphone = 0.88
(i.e. 1 − 0.12 = 0.88)

Probability of both events occurring $= \frac{1}{4} \times 0.88 = 0.22$

The second scenario is that he misses the first train because he is distracted by his smartphone and then waits between 12 and 19 minutes.

Probability of waiting between 12 and 19 minutes:

$$P(12 < X < 19) = \frac{d - c}{b - a} = \frac{19 - 12}{24 - 12} = \frac{7}{12}$$

Probability of being distracted by smartphone = 0.12

Probability of both events occurring $= 0.12 \times \dfrac{7}{12} = 0.07$

Probability of either scenario = 0.22 + 0.07 = 0.29

> Note that the probability of getting on the first train = 0.22 and this was worked out in part (i).

(ii) P(Gets on first train/waits between 9 and 19 mins) $= \dfrac{0.22}{0.29} = 0.7586 \ldots$

3 (a) Mean, $E(X) = \frac{1}{2}(a + b)$

$$= \frac{1}{2}(-6 + 4)$$

$$= -1$$

(b) $P(X \leq 2.4) = \frac{d - c}{b - a} = \frac{2.4 - -6}{4 - -6}$

$$= \frac{8.4}{10}$$

$$= 0.84$$

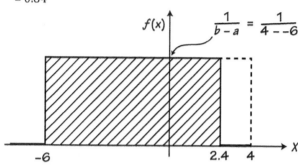

You could also find the probability graphically by working out the area shaded as a fraction of the whole area of the rectangle.

(c) $P(-2 \leq X \leq 3) = \frac{d - c}{b - a} = \frac{3 - -2}{4 - -6}$

$$= \frac{5}{10}$$

$$= \frac{1}{2}$$

(d) $P(-8 \leq X - 3 \leq 3) = P(-5 \leq X \leq 6)$

$P(-5 \leq X \leq 6) = P(-5 \leq X \leq 4)$ as the probability between 4 and 6 is zero.

> Add 3 to each side of the inequality so the inequality just refers to X.

$$P(-5 \leq X \leq 4) = \frac{d - c}{b - a} = \frac{4 - -5}{4 - -6}$$

$$= \frac{9}{10}$$

$$= 0.9$$

4 (a) $X \sim N(5, 0.09)$

$$z = \frac{x - \mu}{\sigma}$$

$$= \frac{4.5 - 5}{0.3}$$

$$= -1.667$$

$$P(Z < -1.667) = 1 - P(Z < 1.667)$$

$$= 1 - 0.95254$$

$$= 0.04746$$

> Note that if you used a calculator to work this out your value for the probability would be slightly different.

(b) $P(X > 4.8) = 0.95$

Looking up a z-value with a probability of 0.95 gives $z = -1.64$

$$z = \frac{x - \mu}{\sigma}$$

$$-1.64 = \frac{4.8 - 5}{\sigma}$$

$$\sigma = 0.122$$

5 (a) Use Inverse Normal on the calculator and enter:

Area: 0.45

σ: 4

μ: 15

Hence $x = 14.50$ (2 d.p.)

A calculator method is used here but you could use tables but you would need to convert to z-values before using the tables.

(b) $P(X > x) = 0.3$

The area representing $P(X > x) = 0.3$ will be at the upper tail.

For $P(X < x)$ we can change this area/probability to $1 - 0.3 = 0.7$.

Use 'Inverse Normal' on the calculator and enter:

Area: 0.7

σ: 4

μ: 15

Hence $x = 17.10$ (2 d.p.)

(c) It is best to draw a sketch of the random variable X first to work out where the area corresponding to a probability of 0.4 lies.

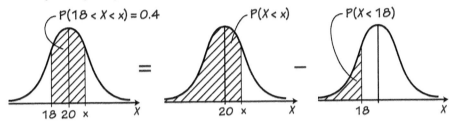

Using the diagram you can see that $P(18 < X < x) = P(X < x) - P(X < 18)$

Now we need to find the z-value for $X = 18$

$$z = \frac{x - \mu}{\sigma}$$

$$z = \frac{18 - 20}{3}$$

$$= -0.67$$

Hence $P(z < -0.67) = 1 - P(z < 0.67) = 1 - 0.74857 = 0.25143$

$0.4 = P(Z < z) - 0.25143$

$P(Z < z) = 0.65143$

Now we use the table backwards to look up the probability and find the corresponding z-value.

The formula $z = \frac{x - \mu}{\sigma}$ can then be used to find the value of x.

From the table, the nearest value to 0.65143 is 0.65173 and this gives a z-value of 0.39.

Using $z = \frac{x - \mu}{\sigma}$

we have: $0.39 = \frac{x - 20}{3}$

Hence $x = 21.17$

Tables have been used for this part of the question but it can be solved using a calculator. If you have used tables, see if you can solve it using a calculator.

Grade boost

Always look back at the question to see if any accuracy is required for your answer. Here the value is to be given to two decimal places.

6 (a) The variable is continuous, and the values are grouped symmetrically around the mean value for weekly household expenditure on food. The distribution is tallest in the middle and tails off equally towards each end.

(b) (i) $P(60 \leq x \leq 70) = 0.26056$

Number of households in this range $= 0.26056 \times 80 = 20.8 = 21$

(ii) $P(x \geq 90) = 1 - P(x < 90)$

$= 1 - 0.95847$

$= 0.04153$

Number of households in this range $= 0.04153 \times 80 = 3.3 = 3$

(c) (i) Using the Normal distribution, the number of households for $60 \leq x \leq 70$ is 21 whereas from the table it is 18. For $x \geq 90$ the Normal distribution predicts 3.3 whereas the table gives a value of 6.

Both of these results mean the Normal distribution predicts different values and may therefore not be the best model.

(ii) The mean predicted by the model needs raising and the tails need to predict a higher probability. Increasing the variance/standard deviation would lower the middle of the curve and lift up the tails, and this would better fit the actual values obtained from the tables.

(d) It may not be suitable as in Northern Ireland weekly household expenditure on food may have a different distribution.

> Note that you could, alternatively, make the point that it could be similar as Northern Ireland is part of the UK as is Wales. Hence it is likely to have a similar socioeconomic status to Wales.

7 (a)

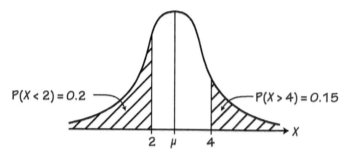

$Z \sim N(0, 1)$ as the standard Normal distribution is being used.

Looking at the higher tail, we have:

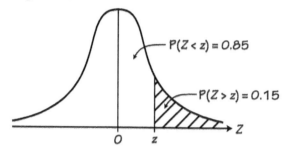

Use the 'Inverse CD' on the calculator with

$\mu = 0$, $\sigma = 1$, and $P(Z < z) = $ Area $= 0.85$

The calculator gives a z-value of 1.0364

When $x = 4$, $z = \dfrac{x - \mu}{\sigma} = \dfrac{4 - \mu}{\sigma}$

Hence $1.0364 = \dfrac{4 - \mu}{\sigma}$

$1.0364\sigma = 4 - \mu$ (1)

Looking at the lower tail, we have:

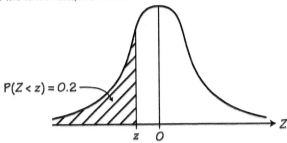

$P(Z < z) = 0.2$

Use the 'Inverse CD' on the calculator with

$$\mu = 0, \ \sigma = 1, \text{ and } P(Z < z) = \text{area} = 0.2$$

Calculator gives a z-value of -0.8416

When $x = 2$, $z = \dfrac{x - \mu}{\sigma} = \dfrac{2 - \mu}{\sigma}$

Hence $-0.8416 = \dfrac{2 - \mu}{\sigma}$

$$-0.8416\,\sigma = 2 - \mu \qquad\qquad\qquad (2)$$

Solving equations (1) and (2) simultaneously

$$\mu = 2.896 \text{ (3 d.p.)}$$
$$\sigma = 1.065 \text{ (3 d.p.)}$$

(b) $P(X > 3)$

Use the 'Normal CD' on the calculator with

     Lower: 3
     Upper: 1×10^{99}
      σ: 1.065
      μ: 2.896

Calculator gives a value for $P(X > 3) = 0.461$ (3 d.p.)

3 Statistical hypothesis testing

1 The null hypothesis is that the mean time is equal to 25 s.
The alternative hypothesis is that the mean time is less than 25 s
Hence we can write, $\mathbf{H_0} : \mu = 25$
          $\mathbf{H_1} : \mu < 25$

Let X be the time to make a coffee.

Assuming $\mathbf{H_0}$, if $X \sim N(\mu, \sigma^2)$ then the sample mean weight, $\overline{X}$, is normally distributed so we can say $\overline{X} \sim N\left(\mu, \dfrac{\sigma^2}{n}\right)$.

Hence $\overline{X} \sim N\left(25, \dfrac{4}{20}\right)$ so $\overline{X} \sim N(25, 0.2)$

and test statistic, $z = \dfrac{\overline{X} - \mu}{\frac{\sigma}{\sqrt{n}}}$

$$= \dfrac{24 - 25}{\frac{2}{\sqrt{20}}}$$

$$= -2.236$$

This is a one-tailed test and the test applies to the lower tail of the distribution.

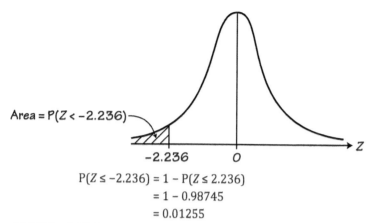

Area = P(Z < –2.236)

$$P(Z \leq -2.236) = 1 - P(Z \leq 2.236)$$
$$= 1 - 0.98745$$
$$= 0.01255$$

Now 0.01255 < 0.05 there is significant evidence at the 5% level of significance to reject H_0 thus suggesting that the new machine has reduced the mean time to make a coffee.

2 (a) The *p*-value for correlation between price of wheat and price of oats of 0.4447 is greater than 0.05. This means there is little evidence to reject the null hypothesis, so the price of wheat and the price of oats do not seem to be correlated.

The product moment correlation coefficient of 0.244 is nearer to 0 than 1 and this also indicates that it is unlikely that there is correlation between the two variables.

> The *p*-value is the probability that you would have found the current result if the correlation coefficient were in fact zero (null hypothesis). If this probability is lower than the conventional 5% ($p < 0.05$) the correlation coefficient is called statistically significant and would give reason to reject the null hypothesis.

(b) $H_0 : \rho = 0$
$H_1 : \rho \neq 0$

> As we are testing for any correlation (i.e. positive or negative) we need to perform a two-tailed test.

We now use Table 9 Critical Value of the Product Moment Correlation Coefficient with $n = 12$, two-tailed test with significance level 5%.

From the table, critical value = 0.5760

The test statistic (i.e. the PMCC) = 0.653

Now 0.653 > 0.5760, so there is evidence to reject H_0

Thus, there is evidence to suggest that there is correlation between the two variables.

(c) The first graph compares two different grains (i.e. wheat and oats) whereas the second graph compares two wheat products, so there is more likely to be correlation in the second example.

3 (a) $H_0 : \rho = 0$
$H_1 : \rho \neq 0$

> Note that we are testing for any correlation as the question does not mention a particular correlation (e.g. positive or negative). Hence, we conduct a two-tailed test.

The test statistic is the correlation coefficient of 0.867

We now refer to the Critical values of the Product Moment Correlation Coefficient table and look up the critical value using the following:

Two tail
5%
$n = 30$

Looking up the critical value, we obtain critical value = ±0.3610

Now as the test statistic (0.867) > critical value (0.3610) there is evidence at the 5% level of significance to reject the null hypothesis H_0.

Hence there is strong evidence to suggest that the correlation coefficient is greater than zero.

(b) The *p*-value for the correlation between '% Fat' and 'Daily calories consumed' is > 0.05 thus indicating that the '% Fat' and 'Daily calories consumed' do not seem to be correlated.

4 (a) $H_0 : \mu = 1.5$

$H_1 : \mu \neq 1.5$

Let X be the length of a screw.

Assuming H_0, if $X \sim N(\mu, \sigma^2)$ then the sample mean weight, $\bar{X}$, is normally distributed so we can say $\bar{X} \sim N\left(\mu, \frac{\sigma^2}{n}\right)$.

Hence $\bar{X} \sim N\left(1.5, \frac{0.05^2}{50}\right)$ so $\bar{X} \sim N(25, 0.00005)$

$$z = \frac{\bar{x} - 1.5}{\frac{0.05}{\sqrt{50}}}$$

$$= \frac{\bar{x} - 1.5}{0.007071}$$

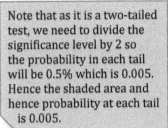

Note that as it is a two-tailed test, we need to divide the significance level by 2 so the probability in each tail will be 0.5% which is 0.005. Hence the shaded area and hence probability at each tail is 0.005.

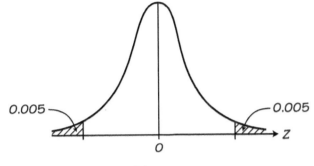

0.005   0.005

Now $P(Z > z) = 0.005$

$P(Z > z) = 1 - P(Z < z)$

Here we look at the upper-tail.

Hence $P(Z < z) = 1 - P(Z > z)$

$= 1 - 0.005$

$= 0.995$

Looking this up in Table 3 the Normal Distribution Table

0.995 gives a z-value of 2.58

As the curve is symmetrical the two z-values are ±2.58.

Hence $2.58 = \dfrac{\bar{x} - 1.5}{0.007071}$ or $-2.58 = \dfrac{\bar{x} - 1.5}{0.007071}$

Solving each of these gives critical values of 1.518 and 1.482

Hence the critical regions are $\bar{X} \leq 1.482$ or $\bar{X} > 1.518$

(b) The mean length of the sample of screws was found to be 1.53 cm, which falls outside the critical region. This means there is sufficient evidence at the 1% level of significance that the null hypothesis should be rejected and that the mean length of the screws has changed, so the machine needs adjusting.

Unit 4 Applied Mathematics B
Section B: Mechanics

4 Kinematics for motion with variable acceleration

1 (a) $v = 64 - \frac{1}{27}t^3$

When the car comes to rest $v = 0$, so we have:

$$0 = 64 - \frac{1}{27}t^3$$

Hence $\quad \frac{1}{27}t^3 = 64$

$$t^3 = 64 \times 27$$

Cube-rooting both sides gives $t = 12$ s

(b) $s = \int_0^{12} v \, dt$

$$= \int_0^{12} \left(64 - \frac{1}{27}t^3\right) dt$$

$$= \left[\left(64t - \frac{1}{108}t^4\right)\right]_0^{12}$$

$$= \left[\left(64 \times 12 - \frac{1}{108}(12)^4\right) - 0\right]$$

$$= 576 \text{ m}$$

> When the car passes P, $t = 0$ and when it passes Q, $t = 12$ so these are used as the limits for the integration.

2 (a) $v = \int a \, dt$

$$= \int (2t + 3) \, dt$$

$$= t^2 + 3t + c$$

When $t = 0$, $v = 10$.

Substituting these values into the above equation to find c, we have:

$$v = t^2 + 3t + c$$
$$10 = 0^2 + 0 + c$$
$$c = 10$$

Hence $\quad v = t^2 + 3t + 10$

When $t = 3$, $\quad v = 3^2 + 3(3) + 10$

$$= 28 \text{ m s}^{-1}$$

(b) $s = \int v \, dt$

$$= \int (t^2 + 3t + 10) \, dt$$

$$= \frac{t^3}{3} + \frac{3t^2}{2} + 10t + c$$

When $t = 0$, $s = 0$.

Substituting these values into the above equation to find c, we have $c = 0$.

Hence $\qquad s = \frac{t^3}{3} + \frac{3t^2}{2} + 10t$

When $t = 3$, $\qquad s = \frac{3^3}{3} + \frac{3(3)^2}{2} + 10(3)$

$$= 9 + 13.5 + 30$$

$$= 52.5 \text{ m}$$

3 (a) $s = \int v \, dt$

$\quad = \int (4t^3 - 6t + 1) \, dt$

$\quad = t^4 - 3t^2 + t + c$

When $t = 0$, $s = 4$ m so $s = t^4 - 3t^2 + t + c$

$\qquad\qquad\qquad\qquad 4 = 0^4 - 3(0)^2 + 0 + c \quad$ giving $c = 4$.

Hence $\qquad\qquad\qquad s = t^4 - 3t^2 + t + 4$

When $t = 2$, $\qquad\qquad s = 2^4 - 3(2)^2 + 2 + 4 = 10$ m

> Remember to include the constant of integration c.
> To find c, substitute $t = 0$ and $s = 4$ into the equation for s.

(b) $a = \dfrac{dv}{dt} = 12t^2 - 6$

When $t = 1$, $a = 12 - 6 = 6$ m s^{-2}

4 (a) $v = 2t^2 - 20t + 32$

$\quad v = 2(t^2 - 10t + 16)$

When $v = 0$, $\qquad 2(t^2 - 10t + 16) = 0$

So $\qquad\qquad\qquad 2(t - 2)(t - 8) = 0$

Hence $\qquad\qquad\qquad\qquad t = 2$ s or 8 s

> We can find the intercept on the y-axis by substituting $t = 0$ into
> $$v = 2t^2 - 20t + 32.$$
> Hence when $t = 0$, $v = 32$ m s^{-1}

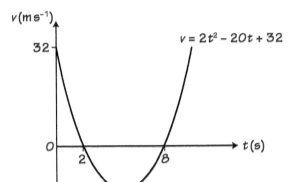

(b) $v = 2t^2 - 20t + 32$

$\quad \dfrac{dv}{dt} = 4t - 20$

At minimum value, $\dfrac{dv}{dt} = 0$, so $4t - 20 = 0$

Hence $t = 5$ s

When $t = 5$ s, $v = 2(5)^2 - 20(5) + 32 = -18$ m s^{-1}

> **Note:** to find the minimum you could have used the symmetry of the curve to give $t = 5$ s.

(c) When $v = 14$ m s^{-1}, $\quad 2t^2 - 20t + 32 = 14$

$\qquad\qquad\qquad\qquad\qquad t^2 - 10t + 9 = 0$

$\qquad\qquad\qquad\qquad\qquad (t - 9)(t - 1) = 0$

$\qquad\qquad\qquad\qquad\qquad\qquad t = 1$ s or 9 s

(d) $s = \int v \, dt$

$\quad = \int (2t^2 - 20t + 32) \, dt$

Now displacement in first 2 seconds $= \displaystyle\int_0^2 \left(2t^2 - 20t + 32\right) dt$

$\qquad\qquad\qquad\qquad\qquad = \left[\dfrac{2t^3}{3} - 10t^2 + 32t\right]_0^2$

$\qquad\qquad\qquad\qquad\qquad = \left[\left(\dfrac{16}{3} - 40 + 64\right) - (0)\right]$

$\qquad\qquad\qquad\qquad\qquad = \dfrac{88}{3}$

> You have to be careful here that you don't integrate using $t = 0$ and 8 as the limits, as that would give the displacement rather than the distance. Note that from the graph you can see that the displacement between the times 2 and 8 seconds is negative as the area is under the x-axis.

Displacement between $t = 2$ s and $t = 8$ s

$$= \int_{2}^{8} \left(2t^2 - 20t + 32\right) dt$$

$$= \left[\frac{2t^3}{3} - 10t^2 + 32t\right]_{2}^{8}$$

$$= \left[\left(\frac{1024}{3} - 640 + 256\right) - \left(\frac{16}{3} - 40 + 64\right)\right]$$

$$= -72$$

(note that this is negative as the area is below the x-axis)

We take the positive value for the distance, (i.e. 72).

Hence total distance travelled $= \frac{88}{3} + 72 = \frac{304}{3}$ m

5 Kinematics for motion using vectors

1　(a)　$\mathbf{F} = 2\mathbf{i} + \mathbf{j} - \mathbf{i} + 2\mathbf{j} + 3\mathbf{i} - \mathbf{j}$

　　　　$= 4\mathbf{i} + 2\mathbf{j}$

　　(b)　Force = mass × acceleration (i.e. $\mathbf{F} = m\mathbf{a}$)

　　　　Hence $\mathbf{a} = \dfrac{\mathbf{F}}{m} = \dfrac{4\mathbf{i} + 2\mathbf{j}}{2} = 2\mathbf{i} + \mathbf{j}$

　　　　Using　$\mathbf{v} = \mathbf{u} + \mathbf{a}t$

　　　　　$\mathbf{v} = (3\mathbf{i} - 7\mathbf{j}) + (2\mathbf{i} + \mathbf{j})t$

　　　　　　$= (3 + 2t)\mathbf{i} + (t - 7)\mathbf{j}$

　　(c)　When $t = 2$, $\mathbf{v} = 7\mathbf{i} - 5\mathbf{j}$

　　　　　　　Speed $= \sqrt{7^2 + (-5)^2}$

　　　　　　　　　$= \sqrt{74}$

　　　　　　　　　$= 8.6$ m s^{-1}

> To find the resultant force (i.e. the single force equivalent to all the other forces) simply add the vectors for the forces together.

> Notice that the acceleration vector does not contain t which means the acceleration is constant, so we can use the equations of motion in subsequent calculations.

2　(a)　If the force is constant, then the acceleration will be constant.

　　　　Force = mass × acceleration (i.e. $\mathbf{F} = m\mathbf{a}$)

　　　　Hence $\mathbf{a} = \dfrac{\mathbf{F}}{m} = \dfrac{5\mathbf{i} + 15\mathbf{j}}{5} = \mathbf{i} + 3\mathbf{j}$

　　　　Using　$\mathbf{v} = \mathbf{u} + \mathbf{a}t$

　　　　　$\mathbf{v} = (-3\mathbf{i} + 5\mathbf{j}) + (\mathbf{i} + 3\mathbf{j}) \times 2$

　　　　　　$= -3\mathbf{i} + 5\mathbf{j} + 2\mathbf{i} + 6\mathbf{j}$

　　　　　　$= -\mathbf{i} + 11\mathbf{j}$

　　(b)　Using　$\mathbf{s} = \mathbf{u}t + \dfrac{1}{2}\mathbf{a}t^2$

　　　　　$\mathbf{s} = (-3\mathbf{i} + 5\mathbf{j})2 + \dfrac{1}{2} \times (\mathbf{i} + 3\mathbf{j})4$

　　　　　　$= -6\mathbf{i} + 10\mathbf{j} + 2\mathbf{i} + 6\mathbf{j}$

　　　　　　$= -4\mathbf{i} + 16\mathbf{j}$

Now at $t = 0$, the position vector is $3\mathbf{i} - 12\mathbf{j}$ so this vector needs to be added to the displacement vector travelled in 2 s to find the total displacement from the origin.

Hence displacement from the origin $= 3\mathbf{i} - 12\mathbf{j} - 4\mathbf{i} + 16\mathbf{j} = -\mathbf{i} + 4\mathbf{j}$

　　　　Distance $= \sqrt{(-1)^2 + (4)^2} = \sqrt{17}$ m

> We know the force is constant because we are told it is in the question.
> Notice that the force does not depend on t and this also tells us the force is constant.

> Notice the question asks for the exact value so do not work out the root as a decimal.

3   Displacement vector $= (7\mathbf{i} + 16\mathbf{j}) - (2\mathbf{i} + 4\mathbf{j})$

$$= 5\mathbf{i} + 12\mathbf{j}$$

Distance between the two points $= \sqrt{5^2 + 12^2}$

$$= \sqrt{169}$$

$$= 13 \text{ m}$$

As the speed is constant, we can use $\text{speed} = \dfrac{\text{distance}}{\text{time}}$

Hence, $\text{time} = \dfrac{\text{distance}}{\text{speed}} = \dfrac{13}{1.3} = 10 \text{ s}$

> To find the displacement vector you subtract the position vector of the starting point from the position vector of the finishing point.

4 $\mathbf{v} = 2 \tan 2t\mathbf{i}$

$\mathbf{a} = \dfrac{\mathrm{d}\mathbf{v}}{\mathrm{d}t} = 4 \sec^2 2t\mathbf{i}$

When $t = \dfrac{\pi}{6}$, $\mathbf{a} = 4 \sec^2 \dfrac{\pi}{3}\mathbf{i}$

Now $\sec^2 \dfrac{\pi}{3} = \dfrac{1}{\cos^2 \frac{\pi}{3}}$ and $\cos \dfrac{\pi}{3} = \dfrac{1}{2}$ so $\cos^2 \dfrac{\pi}{3} = \dfrac{1}{4}$

So $\sec^2 \dfrac{\pi}{3} = \dfrac{1}{\frac{1}{4}} = 4$

Hence $\mathbf{a} = 4 \times 4\mathbf{i}$

$$= 16\mathbf{i} \text{ m s}^{-2}$$

5   (a)   When $t = \dfrac{\pi}{3}$, $\mathbf{r} = \dfrac{\pi}{3} \sin \dfrac{\pi}{3}\mathbf{i} + \dfrac{\pi}{3} \cos \dfrac{\pi}{3}\mathbf{j}$

$$\mathbf{r} = \dfrac{\pi}{3} \times \dfrac{\sqrt{3}}{2}\mathbf{i} + \dfrac{\pi}{3} \times \dfrac{1}{2}\mathbf{j}$$

$$= \dfrac{\sqrt{3}\pi}{6}\mathbf{i} + \dfrac{\pi}{6}\mathbf{j}$$

$$= \dfrac{1}{6}\left(\sqrt{3}\pi\mathbf{i} + \pi\mathbf{j}\right)$$

(b) $\mathbf{r} = t \sin t\mathbf{i} + t \cos^2 t\mathbf{j}$

$$\mathbf{v} = \dfrac{\mathrm{d}\mathbf{r}}{\mathrm{d}t}$$

$$= (\sin t + t \cos t)\mathbf{i} + (\cos t - t \sin t)\mathbf{j}$$

$$\text{Speed}^2 = (\sin t + t \cos t)^2 + (\cos t - t \sin t)^2$$

$$= \sin^2 t + 2t \sin t \cos t + t^2 \cos^2 t + \cos^2 t - 2t \sin t \cos t + t^2 \sin^2 t$$

$$= 1 + t^2$$

Hence,

$$\text{speed} = \sqrt{1 + t^2}$$

> Note that to differentiate $t \sin t\mathbf{i}$ you have to use the Product rule. You also have to use the Product rule to differentiate $t \cos t\mathbf{j}$.

> To simplify this we make use of
> $$\sin^2 t + \cos^2 t = 1$$
> and
> $$t^2 \sin^2 t + t^2 \cos^2 t = t^2(\sin^2 t + \cos^2 t) = t^2$$

6 Types of force, resolving forces and forces in equilibrium

1   Resolving forces in the direction of the 8 N force, we obtain:

$$8 + 6 \cos 60° = R \cos 60°$$

Solving gives $R = 22 \text{ N}$

Resolving forces in the direction of the P N force, we obtain:

$$P = R \sin 60° + 6 \sin 60°$$

$$= 22 \sin 60° + 6 \sin 60°$$

Solving gives $P = 14\sqrt{3} \text{ N}$

> Note that the exact value is required so we do not give the answer as a decimal.

2 (a)

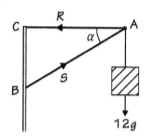

R is the tension,
S is the thrust

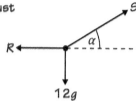

> R is the tension in rod AC
> and S is the thrust in rod AB.

The forces acting at point A are in equilibrium.
Resolving vertically, we have
$$S \sin \alpha = 12g$$
$$S \times 0.6 = 12 \times 9.8$$
Thrust in AB, $\quad S = 196 \, \text{N}$

> Note that if, $\sin \alpha = 0.6$,
> $\cos \alpha = 0.8$

(b) Resolving horizontally, we have
$$S \cos \alpha = R$$
Tension in AC, $\quad R = 156.8 \, \text{N}$

3 Resolving in the direction of the 20 N force, we obtain
$$150 \sin 30° = 20 + Q \cos \theta$$
$$Q \cos \theta = 75 - 20$$
$$Q \cos \theta = 55 \, \text{N} \qquad\qquad (1)$$
Resolving in the direction at right-angles to the 20 N force, we obtain
$$150 \cos 30° + Q \sin \theta = 140$$
$$Q \sin \theta = 10.096 \qquad\qquad (2)$$
Squaring equations (1) and (2) and then adding them together, we obtain
$$Q^2 \sin^2 \theta + Q^2 \cos^2 \theta = (10.096)^2 + 55^2$$
$$Q^2(\sin^2 \theta + \cos^2 \theta) = 101.933 + 3025$$
$$Q^2 = 3126.933$$
$$Q = 55.92 \, \text{N}$$
Dividing equation (2) by equation (1) we obtain
$$\tan \theta = \frac{10.096}{55}$$
$$\theta = 10.40°$$

4 (a) A tension in a rod is a force produced in a rod when the rod is being pulled. A thrust in a rod is when the rod is being compressed.

(b)

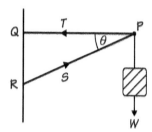

T = Tension
S = Thrust
W = Weight

(c) The forces acting at point P are in equilibrium.
Resolving vertically, we have:
$$S \sin \theta = 2g$$
As $\sin \theta = 0.5$ and $g = 9.8$
$$0.5S = 19.6$$
$$S = 39.2 \, \text{N}$$
Thrust in rod PR = 39.2 N

(d) Resolving horizontally, we have:
$$T = S \cos \theta$$
$$= 39.2 \cos 30°$$
$$= 33.9 \, N$$
Tension in rod PQ = 33.9 N

(e) So that the weights of the rods could be neglected.

7 Forces and Newton's laws

Resolving vertically, we obtain: $R = 12g$
$$= 12 \times 9.8$$
Limiting friction, $F = \mu R$
$$= 0.8 \times 12 \times 9.8$$
$$= 94.08 \, N$$

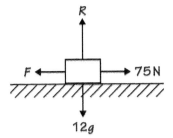

This is the maximum frictional force, so the friction can be any value less than or equal to this value.

The tractive force of 75 N is less than this value, so the frictional force will match this value and the particle will remain at rest.

Hence frictional force = 75 N

2 (a)

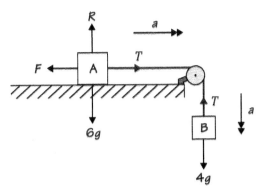

Resolving vertically for A, we have
$$R = 6g$$
Limiting friction, $F = \mu R = 0.4 \times 6g = 2.4g$
Applying Newton's 2nd law to particle A, we have
$$ma = T - F$$
$$6a = T - 2.4g \qquad (1)$$
Applying Newton's 2nd law to particle B, we have
$$ma = 4g - T$$
$$4a = 4g - T \qquad (2)$$
Adding equations (1) and (2), we obtain
$$10a = 1.6g$$
$$10a = 1.6 \times 9.8$$
Hence $\qquad a = 1.57 \, m \, s^{-2}$
Substituting this value of a into equation (1), we obtain
$$6 \times 1.57 = T - (2.4 \times 9.8)$$
Hence $\qquad T = 32.93 \, N$

(b) The tension is constant in the string.

3 (a)

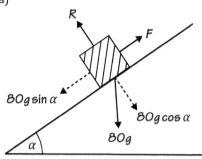

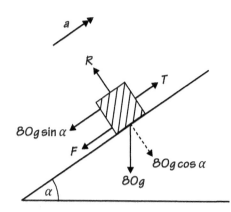

If $\sin \alpha = \frac{3}{5}$, $\cos \alpha = \frac{4}{5}$

Resolving at right angles to the plane, we have
$$R = mg \cos \alpha = 80 \times 9.8 \times 0.8 = 627.2 \, \text{N}$$

(b) As there is no acceleration along the slope the forces parallel to the slope are in equilibrium (i.e. frictional force is equal to the component of the weight down the slope).

Resolving parallel to the slope, we obtain
$$F = 80g \sin \alpha = 80 \times 9.8 \times \frac{3}{5} = 470.4 \, \text{N}$$

On the point of slipping, $F = \mu R$

Hence,
$$\mu = \frac{F}{R} = \frac{470.4}{627.2} = 0.75$$

This is limiting friction.

(c)

Notice that the forces at right angles to the plane have not altered,
so $R = 627.2 \, \text{N}$.

Also, the magnitude of the frictional force will not change but its direction is now down the slope.

Hence $F = 470.4 \, \text{N}$.

Applying Newton's 2nd law parallel to the slope, we have
$$ma = T - 80g \sin \alpha - F$$

$$80 \times 0.7 = T - \left(80 \times 9.8 \times \frac{3}{5} \right) - 470.4$$

giving $T = 996.8 \, \text{N}$

4 (a) Resultant force in the direction $OX = 8 \cos 60° + 6\sqrt{3} \cos 30°$
$$= 13\,\text{N}$$

Resultant force in the direction $OY = 8 \sin 60° - 6\sqrt{3} \sin 30°$
$$= 1.73\,\text{N}$$

We can now draw a diagram of these forces:

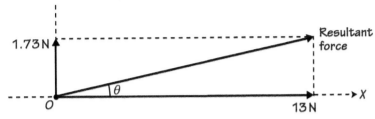

Using Pythagoras' theorem we obtain:
$$R^2 = (1.73)^2 + (13)^2$$
$$R = 13.11\,\text{N (2 d.p.)}$$

$$\theta = \tan^{-1}\left(\frac{1.73}{13}\right)$$
$$= 7.58°\,\text{(2 d.p.)} \quad \text{to the line OX in the direction shown on the diagram.}$$

(b) $a = \dfrac{F}{m}$

$$= \frac{13.11}{2}$$
$$= 6.56\,\text{m s}^{-2}$$

5 (a)

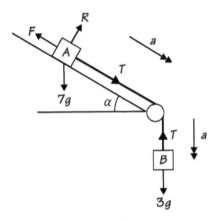

Resolving perpendicular to plane, we obtain:
$$R = 7g \cos \alpha$$
Now $F = \mu R,$ so $F = \mu 7g \cos \alpha$

$$F = 0.6 \times 7 \times 9.8 \times \frac{4}{5}$$
$$= 32.928\,\text{N}$$

> Note that as $\tan \alpha = \frac{3}{4}$ the hypotenuse of the right-angled triangle would be of length 5.
> Hence $\cos \alpha = \frac{4}{5}$ and $\sin \alpha = \frac{3}{5}$.

(b) Applying Newton's 2nd law of motion to A, we obtain:
$$T + 7g \sin \alpha - F = 7a$$
$$T + 41.16 - 32.928 = 7a$$
$$T + 8.232 = 7a \tag{1}$$

> Here we have taken the direction to the right as the positive direction.

Applying Newton's 2nd law of motion to B, we obtain:
$$3g - T = 3a$$
$$3g + 8.232 = 10a \tag{2}$$
$$a = 3.76\,\text{m s}^{-2}$$
$$T = 18.11\,\text{N}$$

> Here we have taken the direction downwards as the positive direction.

> Equations (1) and (2) are solved simultaneously by adding them together to eliminate T.

> a is substituted back into either (1) or (2) to find T.

6 (a)

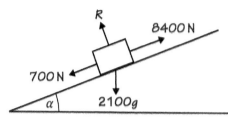

We can consider the car and trailer as a single object of mass 2100 kg. Drawing all the forces acting, we obtain the above diagram.

Applying Newton's 2nd law, we obtain:

$$8400 - 700 - 2100g \sin \alpha = 2100a$$
$$8400 - 700 - 5762.4 = 2100a$$

Solving gives the acceleration of the car, $a = \frac{346}{375} = 0.923 \text{ m s}^{-2}$ (up the slope)

> Here we have taken the upward direction as positive.

(b)

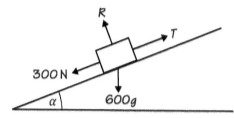

Applying Newton's 2nd law to the trailer only, we obtain:

$$T - 300 - 600g \sin \alpha = 600a$$
$$T - 300 - 600 \times 9.8 \times \frac{7}{25} = 600 \times \frac{346}{375}$$
$$T = 2500 \text{ N}$$

> Taking the upward direction as positive.

8 Projectile motion

1 (a)

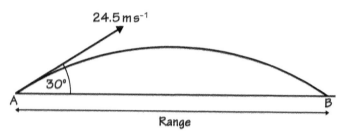

Considering the vertical component and letting the upward direction be positive, we have:

$$s = ut + \frac{1}{2}at^2$$

The initial vertical velocity, $u = 24.5 \sin 30° = 12.25 \text{ m s}^{-1}$

At the range, $s = 0$, so:

$$0 = 12.25t - \frac{1}{2} \times 9.8 \times t^2$$
$$= t(12.25 - 4.9t)$$

Solving gives, $t = 2.5 \text{ s}$

> Note there is another solution to this equation which is $t = 0$ which we ignore.

Range = horizontal velocity × time of flight
$$= 24.5 \cos 30° \times 2.5$$
$$= 53 \text{ m}$$

> Note there is no acceleration in the horizontal direction so the velocity in this direction stays constant.

(b)

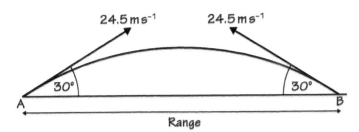

On projection, both objects have the same vertical and horizontal velocities.

Vertical velocity = 12.25 m s^{-1}

Horizontal velocity = 24.5 cos 30° = 12.25√3 m s^{-1}

If the particle from A is in the air t s before colliding, the particle from B will be in the air $(t - 1)$ s before colliding.

Horizontal distance travelled by A before colliding = 12.25√3t

Horizontal distance travelled by B before colliding = 12.25√3$(t - 1)$

These two distances will add to give the range.

Hence,   12.25√3t + 12.25√3$(t - 1)$ = 53

Dividing both sides by 12.25√3t we have:

$$t + t - 1 = 2.5$$

Hence $t = 1.75$ s

In this time, the particle projected from A will have travelled a vertical distance

$$s = ut + \frac{1}{2}at^2$$

$$s = 12.25 \times 1.75 - \frac{1}{2} \times 9.8 \times (1.75)^2 = 6.4 \text{ m}$$

Hence, they collide at a height of 6.4 m above the ground.

2 (a) Initial horizontal velocity = 35 cos α

= 35 × 0.6

= 21 m s^{-1}

Initial vertical velocity = 35 sin α

= 35 × 0.8

= 28 m s^{-1}

Using $s = ut + \frac{1}{2}at^2$ in the vertical direction, we obtain

$$0 = 28t - \frac{1}{2} \times 9.8t^2$$

$$t(28 - 4.9t) = 0$$

Solving, gives $t = 0, \frac{40}{7}$

We ignore the solution t = 0 s as this is when the ball was first projected.

Hence $t = \frac{40}{7}$ s

Horizontal velocity is constant at 21 m s^{-1}.

Horizontal distance travelled by the ball = velocity × time

$$= \frac{40}{7} \times 21$$

$$= 120 \text{ m}$$

As this horizontal distance is greater than 100 m, the ball will not fall into the lake.

(b) Time taken to reach the tree = $\frac{\text{distance}}{\text{horizontal velocity}} = \frac{17.5}{21} = \frac{5}{6}$

Using $v = u + at$ in the vertical direction, we obtain

$$v = 28 - 9.8 \times \frac{5}{6}$$

$$= \frac{119}{6} = 19.8333 \text{ m s}^{-1}$$

> **Grade boost**
> Always think about what you have found in the previous part to a question. Often it acts as a guide to what to do for the next part. Here we found the range, so we should approach this question by looking at the horizontal distance travelled by each object.

$$\text{Speed} = \sqrt{\left(\frac{119}{6}\right)^2 + (21)^2}$$

$$= 28.89 \text{ m s}^{-1}$$

$$\text{Angle } \theta = \tan^{-1}\left(\frac{119}{6 \times 21}\right)$$

$$= 43.36°$$

3 (a)

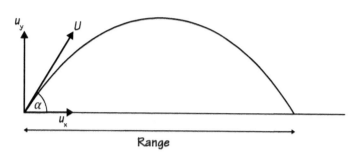

To find the time of flight, we consider the vertical motion and equate the vertical displacement to zero. We will consider the upward direction as the positive direction for the motion.

Before you start the answer, think about which direction you will consider as the positive direction.

$$s = ut + \frac{1}{2}at^2$$

$$s = U \sin \alpha t - \frac{1}{2}gt^2$$

As $s = 0$, $\qquad 0 = U \sin \alpha t - \frac{1}{2}gt^2$

Hence $t(U \sin \alpha - \frac{1}{2}gt) = 0$ so either $t = 0$ or $U \sin \alpha - \frac{1}{2}gt = 0$.

The solution $t = 0$ is when the particle is projected, so we use the other solution.

Hence, time of flight, $t = \dfrac{2U \sin \alpha}{g}$

As the horizontal velocity remains constant during the time of flight, to find the range we use:

Range = horizontal velocity × time of flight

Now horizontal velocity = $U \cos \alpha$ and time of flight = $\dfrac{2U \sin \alpha}{g}$

so substituting these into the above formula, we obtain:

$$\text{Range} = U \cos \alpha \frac{2U \sin \alpha}{g}$$

$$= \frac{2U^2 \sin \alpha \cos \alpha}{g}$$

Now $2 \sin \alpha \cos \alpha \equiv \sin 2\alpha$

$$\text{Range} = \frac{U^2 \sin 2\alpha}{g}$$

Remember that this is the double angle formula.

(b) $U = 42$ m s^{-1} and $R = 90$ m

$$\text{Range} = \frac{U^2 \sin 2\alpha}{g}$$

$$90 = \frac{42^2 \sin 2\alpha}{9.8}$$

$$\sin 2\alpha = \frac{1}{2}$$

Hence
$$2\alpha = 30° \text{ or } 150°$$
$$\alpha = 15° \text{ or } 75°$$

Time of flight is given by $t = \dfrac{2U \sin \alpha}{g}$

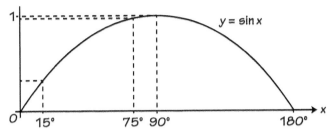

You can see that the value for sin 75° is greater than the value for sin 15°, so the time of flight will be greater for $\alpha = 75°$

(c)   The only force acting is gravity (i.e. no frictional forces/air resistance).

> You could alternatively just put the angles into the calculator to find which angle gives the larger value of sin α.

9 Differential equations

1   (a)   Resistance, $R \propto v$, so $R = kv$.

Now, when $v = 0.2$, $R = 0.08$, so $0.08 = k \times 0.2$, giving $k = 0.4$

Hence $R = 0.4v$

Taking the upward direction as positive and applying Newton's 2nd law of motion we have:

$$m\frac{dv}{dt} = -mg - R$$

$$0.5\frac{dv}{dt} = -9.8 \times 0.5 - 0.4v$$

Dividing both sides by 0.5 we obtain:

$$\frac{dv}{dt} = -9.8 - 0.8v$$

> We have taken upward as the positive direction as this is the direction of motion. The weight and resistive force are both acting downwards and are therefore negative.

(b)   Separating variables and integrating, we obtain:

$$\int \frac{dv}{(-9.8 - 0.8v)} = \int dt$$

Dividing both sides by –1 we have

$$\int \frac{dv}{(9.8 + 0.8v)} = -\int dt$$

$$\frac{1}{0.8} \int \frac{0.8}{(9.8 + 0.8v)}dv = -\int dt$$

$$1.25 \ln (9.8 + 0.8v) = -t + c$$

Now when $t = 0$, $v = 24$

so $1.25 \ln (9.8 + 0.8 \times 24) = c$

Hence $c = 4.2091$

$$1.25 \ln (9.8 + 0.8v) = -t + 4.2091$$

Hence $t = 4.2091 - 1.25 \ln (9.8 + 0.8v)$

> The derivative of the denominator on the left-hand side of the equation is –0.8 so we need to multiply the top and the bottom by 0.8 which means the integral will be 0.8 times the ln of the denominator.

(c)   When $v = 0$, $t = 4.2091 - 1.25 \ln 9.8$
                              $= 1.36 \text{ s (2 d.p.)}$

> At the highest point, the velocity will equal zero.

2 (a) (i) Taking the upward direction as positive and applying Newton's 2nd law of motion, we obtain:

$$m\frac{dv}{dt} = -mg - 3v$$

$$6\frac{dv}{dt} = -6 \times 9.8 - 3v$$

> Dividing both sides of this equation by 3.

$$2\frac{dv}{dt} = -19.6 - v$$

(ii) Separating variables and integrating, we obtain:

$$2\int \frac{dv}{(19.6 + v)} = -\int dt$$

$$2\ln(19.6 + v) = -t + c$$

When $t = 0$, $v = 24.5$ so:

$$c = 2\ln 44.1$$

Hence, $\qquad 2\ln(19.6 + v) = -t + 2\ln 44.1$

$$-t = 2\ln\left(\frac{19.6 + v}{44.1}\right)$$

$$\frac{-t}{2} = \ln\left(\frac{19.6 + v}{44.1}\right)$$

Taking exponentials of both sides, we obtain:

$$e^{-\frac{t}{2}} = \frac{19.6 + v}{44.1}$$

$$v = 44.1\,e^{-\frac{t}{2}} - 19.6$$

(b) At the maximum height, $v = 0$, so $\quad -t = 2\ln\left(\frac{19.6 + v}{44.1}\right)$

$$-t = 2\ln\left(\frac{19.6}{44.1}\right)$$

$$t = 1.62\ s$$

(c) $v = \dfrac{dx}{dt}$

$$x = \int v\,dt$$

$$= \int\left(44.1e^{-\frac{t}{2}} - 19.6\right) dt$$

$$= -88.2e^{-\frac{t}{2}} - 19.6t + c$$

When $t = 0$, $x = 0$ giving $c = 88.2$

$$x = 88.2 - 88.2e^{-\frac{t}{2}} - 19.6t$$

3 (a) Taking the direction of the 1400 N force as positive and applying Newton's 2nd law of motion, we obtain:

$$m\frac{dv}{dt} = 1400 - R \quad \text{where } R \text{ is the resistive force}$$

Now $R \propto t$ so $R = kt$

Hence, $\qquad m\dfrac{dv}{dt} = 1400 - kt$

$$20\frac{dv}{dt} = 1400 - kt$$

$$\frac{dv}{dt} = 70 - \frac{kt}{20}$$

When $t = 4$, $a = -2$ and $\dfrac{dv}{dt} = -2$

Hence $\qquad -2 = 70 - \dfrac{k \times 4}{20}$

$\qquad\qquad \dfrac{k}{5} = 72$

$\qquad\qquad k = 360$

So, $\qquad \dfrac{dv}{dt} = 70 - \dfrac{360t}{20}$

$\qquad\qquad \dfrac{dv}{dt} = 70 - 18t$

(b) Separating variables and integrating, we obtain:

$$\int dv = \int (70 - 18t)\, dt$$
$$v = 70t - 9t^2 + c$$

When $t = 4$, $v = 24$ so $\qquad 24 = 70 \times 4 - 9 \times 4^2 + c$

$\qquad\qquad\qquad\qquad\qquad 24 = 280 - 144 + c$

$\qquad\qquad\qquad\qquad\qquad\quad c = -112$

Hence $\qquad v = 70t - 9t^2 - 112$

> We now need to look at the question for the information that will enable the value of c to be found. We need a value for v and a value for t.

(c) When $t = 5$, $\quad v = 70 \times 5 - 9 \times 5^2 - 112$

$\qquad\qquad\qquad = 13\ \text{m s}^{-1}$

4 (a) $\dfrac{dx}{dt} \propto -x$ so $\dfrac{dx}{dt} = -kx$

Separating variables and integrating, we obtain:

$$\int \dfrac{dx}{x} = -k \int dt$$

$$\ln x = -kt + c$$

> The minus sign is included here because x decreases with time.

When $t = 0$, $x = 2$

Hence, $\ln 2 = 0 + c$, so $c = \ln 2$ and $\ln x = -kt + \ln 2$

$\qquad\qquad\qquad\qquad\qquad \ln x - \ln 2 = -kt$

$\qquad\qquad\qquad\qquad\qquad\quad \ln \dfrac{x}{2} = -kt$

When $t = 1$, $x = 1.6$ so $\qquad \ln \dfrac{1.6}{2} = -k$

> Note that $\ln A - \ln B = \ln \dfrac{A}{B}$

Hence $\qquad\qquad\qquad\qquad k = -\ln 0.8$

$\qquad\qquad\qquad\qquad\qquad k = 0.2231$

Hence $\qquad\qquad\qquad\qquad \ln \dfrac{x}{2} = -0.2231t$

Taking exponentials of both sides $\quad \dfrac{x}{2} = e^{-0.2231t}$ so $x = 2e^{-0.2231t}$

When $t = 3$, $\qquad\qquad\qquad x = 2e^{-0.2231 \times 3}$

$\qquad\qquad\qquad\qquad\qquad = 1.024\ \text{mg per litre}$

(b) $x = 2e^{-0.2231t}$

$\qquad\qquad 0.5 = 2e^{-0.2231t}$

$\qquad e^{-0.2231t} = 0.25$

Taking $\ln$ of both sides, we obtain:

$\qquad -0.2231t = \ln 0.25$

$\qquad\qquad\qquad t = 6.21\ \text{hours or } 6\,\text{h } 13\,\text{min}$

Sample Test Paper Unit 3
Pure Mathematics B

1 Assuming there is a real value of x for which $\sin x + \cos x < 1$.
$$\sin x + \cos x < 1$$
Squaring both sides, we obtain $\sin^2 x + 2\sin x \cos x + \cos^2 x < 1$
Now $\sin^2 x + \cos^2 x = 1$ and $2\sin x \cos x = \sin 2x$, so
$$1 + \sin 2x < 1$$
$$\sin 2x < 0$$
In the domain $0 \le x \le \frac{\pi}{2}$, $\sin 2x$ is positive or zero (i.e. not negative) which is a contradiction.

Hence the statement $\sin x + \cos x \ge 1$ in the domain $0 \le x \le \frac{\pi}{2}$ is true.

2 $|2x + 1| = 3|x - 2|$
Squaring both sides, we obtain:
$$4x^2 + 4x + 1 = 9(x^2 - 4x + 4)$$
$$4x^2 + 4x + 1 = 9x^2 - 36x + 36$$
$$5x^2 - 40x + 35 = 0$$
$$x^2 - 8x + 7 = 0$$
$$(x - 1)(x - 7) = 0$$
$$x = 1 \text{ or } 7$$

> Always look to see if there is a common factor which can be divided into all the terms. In this case the common factor is 5.

3 $S_n = \dfrac{n}{2}\Big[2a + (n - 1)d\Big]$
$$1084 = 4(2a + 7d)$$
$$271 = 2a + 7d \tag{1}$$
$$t_n = a + (n - 1)d$$
$$t_3 = a + (3 - 1)d$$
$$206 = a + 2d \tag{2}$$
Multiplying equation (2) by 2 we obtain:
$$412 = 2a + 4d \tag{3}$$
(1) − (3) gives $\qquad -141 = 3d$
Hence, $\qquad\qquad d = -47$
Substituting this value into (1)
$$271 = 2a - 7(47)$$
$$a = 300$$
Hence first term = 300 and common difference = −47

4 The following approximations can be used:
$$\sin \theta \approx \theta \qquad \cos \theta \approx 1 - \frac{\theta^2}{2}$$
$$12 \sin x - 2 \cos x = 8x^2$$
$$12x - 2\left(1 - \frac{x^2}{2}\right) = 8x^2$$
$$12x - 2 + x^2 = 8x^2$$
$$7x^2 - 12x + 2 = 0$$
Using $\qquad x = \dfrac{-b \pm \sqrt{b^2 - 4ac}}{2a}$
$$= \frac{12 \pm \sqrt{144 - 56}}{14}$$

$$= \frac{12 \pm \sqrt{88}}{14}$$

$$= \frac{12 - \sqrt{88}}{14}$$

$$= \frac{12 - 9.3808}{14}$$

$$= 0.19 \text{ rad (2 d.p.)}$$

5 (a) $f(x) = e^{5-\frac{x}{2}} + 6$

Note that the other root would be too large, so it is neglected.

$$y = e^{5-\frac{x}{2}} + 6$$

$$y - 6 = e^{5-\frac{x}{2}}$$

$$\ln(y - 6) = 5 - \frac{x}{2}$$

$$x = 2(5 - \ln(y - 6))$$

$$f^{-1}(x) = 2(5 - \ln(x - 6))$$

Subtract 6 from both sides so that the exponential function is on its own on one side.

Take ln of both sides to remove the exponential function.

(b) Domain of f^{-1} = range of f

As $x \to -\infty$, $e^{5-\frac{x}{2}} + 6 \to \infty$

When $x = 10$, $e^{5-\frac{x}{2}} + 6 = e^{5-\frac{10}{2}} + 6 = e^0 + 6 = 7$

$$R(f) = [7, \infty)$$

Hence $$D(f^{-1}) = [7, \infty)$$

6 Let $\dfrac{11 + x - x^2}{(x + 1)(x - 2)^2} = \dfrac{A}{x + 1} + \dfrac{B}{x - 2} + \dfrac{C}{(x - 2)^2}$

$$11 + x - x^2 = A(x - 2)^2 + B(x - 2)(x + 1) + C(x + 1)$$

Let $x = 2$,   so $9 = 3C$ and $C = 3$

Let $x = -1$, so $9 = 9A$ and $A = 1$

Let $x = 0$,   so $11 = 4A - 2B + C$ and $11 = 4 - 2B + 3$ so $B = -2$

$$f(x) = \frac{1}{x + 1} - \frac{2}{x - 2} + \frac{3}{(x - 2)^2}$$

$$= (x + 1)^{-1} - 2(x - 2)^{-1} + 3(x - 2)^{-2}$$

$$f'(x) = -(x + 1)^{-2} + 2(x - 2)^{-2} - 6(x - 2)^{-3}$$

$$= -\frac{1}{(x + 1)^2} + \frac{2}{(x - 2)^2} - \frac{6}{(x - 2)^3}$$

When $x = 0$, $f'(x) = -1 + \frac{1}{2} + \frac{3}{4}$

$$= \frac{1}{4}$$

7 (a) Differentiating implicitly and using the Product rule:

$$3x^2 3y^2 \frac{dy}{dx} + y^3 6x = 0$$

$$9x^2 y^2 \frac{dy}{dx} + 6xy^3 = 0$$

$$\frac{dy}{dx} = \frac{-6xy^3}{9x^2 y^2}$$

$$= -\frac{2y}{3x}$$

151

(b)   At P, $\dfrac{dy}{dx} = -\dfrac{4}{9}$,   so $-\dfrac{2y}{3x} = -\dfrac{4}{9}$

   Hence $18y = 12x$

   $$3y = 2x$$

   $$x = \frac{3y}{2}$$

   Substituting $x = \dfrac{3y}{2}$ into the equation for C, we obtain:

   $$3\left(\frac{3y}{2}\right)^2 y^3 = 216$$

   $$\frac{27y^5}{4} = 216$$

   $$y^5 = 32$$

   $$y = 2$$

   When $y = 2$, $x = \dfrac{3(2)}{2} = 3$

   Hence $a = 3$ and $b = 2$

8   (a)   Acute angle $AOB = 2\pi - \dfrac{5\pi}{3} = \dfrac{\pi}{3}$

   Length of minor arc AB $= r\theta$

   $$= 9 \times \frac{\pi}{3}$$

   $$= 3\pi \text{ cm}$$

   (b)   Shaded area $= \dfrac{1}{2}r^2\theta$

   $$= \frac{1}{2} \times 9^2 \times \frac{\pi}{3}$$

   $$= 13.5\pi$$

   $$= 42.41 \text{ cm}^2 \text{ (2 d.p.)}$$

9 $\displaystyle\int_0^1 \sqrt{4 - 4x^2}\, dx = \int_0^1 2\sqrt{1 - x^2}\, dx$

   Let $x = \sin\theta$ so $\dfrac{dx}{d\theta} = \cos\theta$ and $dx = \cos\theta\, d\theta$

   When $x = 1$, $\sin\theta = 1$ so $\theta = \dfrac{\pi}{2}$

   When $x = 0$, $\sin\theta = 0$ so $\theta = 0$

> As the variable has been changed from x to θ, the limits are changed to apply to θ.

   $$I = \int_0^{\frac{\pi}{2}} 2\sqrt{1 - \sin^2\theta}\,\cos\theta\, d\theta$$

> $1 - \sin^2\theta = \cos^2\theta$

   $$= \int_0^{\frac{\pi}{2}} 2\sqrt{\cos^2\theta}\,\cos\theta\, d\theta$$

   $$= \int_0^{\frac{\pi}{2}} 2\cos^2\theta\, d\theta$$

   $$= \int_0^{\frac{\pi}{2}} (1 + \cos 2\theta)\, d\theta$$

> Here we use
> $2\cos^2\theta = 1 + \cos 2\theta$

   $$= \left[\theta + \frac{\sin 2\theta}{2}\right]_0^{\frac{\pi}{2}}$$

   $$= \left[\left(\frac{\pi}{2} + \frac{\sin \pi}{2}\right) - (0 + 0)\right]$$

   $$= \frac{\pi}{2}$$

⑩ $\dfrac{dx}{dt} \propto x$

$\dfrac{dx}{dt} = kx$

Separating variables and integrating, we obtain:

$$\int \frac{1}{x}\,dx = k\int dt$$

$$\ln x = kt + c$$

When $t = 0$, $x = 2000$ so $\ln 2000 = 0 + c$

Hence $\qquad\qquad\qquad c = \ln 2000$

Substituting c back into the equation, we obtain:

$$\ln x = kt + \ln 2000$$

When $t = 5$, $x = 2350$ so $\ln 2350 = 5k + \ln 2000$

Hence $\qquad\qquad\qquad k = \dfrac{1}{5}\ln\left(\dfrac{2350}{2000}\right)$

Substituting this value of k back into the equation, we obtain:

$$\ln x = t \times \frac{1}{5}\ln\left(\frac{2350}{2000}\right) + \ln 2000$$

When $t = 9$, $\qquad \ln x = 9 \times \dfrac{1}{5}\ln\left(\dfrac{2350}{2000}\right) + \ln 2000$

$$= 7.8912$$

Taking exponentials of both sides, we obtain:

$$x = e^{7.8912}$$

$$= 2674$$

Population in 2019 = 2674

⑪ (a) $\qquad\quad 2\operatorname{cosec}^2\theta + \cot\theta = 8$

$\qquad\qquad\quad 2(1 + \cot^2\theta) + \cot\theta = 8$

$\qquad\qquad\quad 2 + 2\cot^2\theta + \cot\theta = 8$

$\qquad\qquad\quad 2\cot^2\theta + \cot\theta - 6 = 0$

$\qquad\qquad\quad (2\cot^2\theta - 3)(\cot\theta + 2) = 0$

$\qquad\quad \cot\theta = \dfrac{3}{2} \quad$ or $\quad \cot\theta = -2$

$\qquad\quad \tan\theta = \dfrac{2}{3} \quad$ or $\quad \tan\theta = -\dfrac{1}{2}$

$\qquad\quad \theta = \tan^{-1}\left(\dfrac{2}{3}\right) = 33.69°,\ 213.69°$

$\qquad\quad \theta = \tan^{-1}\left(-\dfrac{1}{2}\right) = 153.43°,\ 333.43°$

> $\operatorname{cosec}^2 x = 1 + \cot^2 x$

(b) (i) $\quad 5\cos\theta + 12\sin\theta = R\cos(\theta - \alpha)$

$\qquad\qquad\qquad\qquad\quad = R\cos\theta\cos\alpha - R\sin\theta\sin\alpha$

$\qquad R\cos\alpha = 5 \ $ and $\ R\sin\alpha = 12$

$\qquad\qquad \dfrac{R\sin\alpha}{R\cos\alpha} = \tan\alpha = \dfrac{12}{5}$

$\qquad \tan\alpha = \dfrac{12}{5} \ $ so $\ \alpha = 67.4°$ (1 d.p.)

$\qquad\qquad\qquad R = \sqrt{5^2 + 12^2} = \sqrt{169} = 13$

Hence $\ 5\cos\theta + 12\sin\theta = 13\cos(\theta - 67.4°)$

(ii) $\dfrac{1}{5\cos\theta + 12\sin\theta + 25} = \dfrac{1}{13\cos(\theta - 67.4°) + 25}$

The least value occurs when the denominator is greatest. The greatest value of the cos function is +1.

Hence least value is $\dfrac{1}{13+24} = \dfrac{1}{37}$

This value occurs when $\cos(\theta - 67.4°) = 1$

Hence $\qquad\qquad\qquad\qquad \theta - 67.4° = 0$

So $\qquad\qquad\qquad\qquad\qquad \theta = 67.4°$

12 (a) $x = \ln t$ so $\dfrac{dx}{dt} = \dfrac{1}{t}$

$y = 6t^2$ so $\dfrac{dy}{dt} = 12t$

Hence, $\qquad \dfrac{dy}{dx} = \dfrac{dy}{dt} \times \dfrac{dt}{dx}$

$\qquad\qquad\qquad = 12t \times t$

$\qquad\qquad\qquad = 12t^2$

(b) $\dfrac{d}{dt}\left(\dfrac{dy}{dx}\right) = 24t$

$\qquad\qquad \dfrac{d^2y}{dx^2} = \dfrac{d}{dt}\left(\dfrac{dy}{dx}\right)\dfrac{dt}{dx}$

$\qquad\qquad\qquad = 24t \times t$

$\qquad\qquad\qquad = 24t^2$

Now $\qquad \dfrac{d^2y}{dx^2} = 6$

So $\qquad 24t^2 = 6$

$\qquad\qquad t^2 = \dfrac{1}{4}$

$\qquad\qquad t = \pm\dfrac{1}{2}$

Now the question says $t > 0$, so t cannot be negative.

Hence $\qquad t = \dfrac{1}{2}$

> Always check with the question if there is more than one answer to see which answer or answers are allowed.

13 (a)

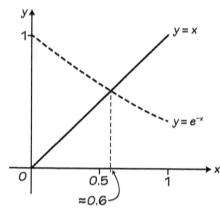

> When $x = 1$, $y = e^{-1} = 0.37$ and when $x = 0$, $y = 1$.

(b) (i) There is only one point of intersection in the given domain, hence only one root of the equation $e^{-x} = x$.

(ii) Estimate of root is 0.6 (1 d.p.)

(iii) $x_0 = 0.6$

$$x_1 = e^{-0.6} = 0.5488116361$$
$$x_2 = e^{-0.5488116361} = 0.5776358443$$
$$x_3 = e^{-0.5776358443} = 0.5612236194$$
$$x_4 = e^{-0.5612236194} = 0.5705105488$$
$$x_5 = e^{-0.5705105488} = 0.5652367841$$
$$x_6 = e^{-0.5652367841} = 0.5682255841$$

The last three answers give a constant value of 0.57 correct to two decimal places.

Hence root = 0.57 (2 d.p.)

Sample Test Paper Unit 3
Unit 4 Applied Mathematics B

Section A – Statistics

1 (a) Let A = the event having the flu jab

Let B = the event catching flu

> Always define the letters you are going to use.

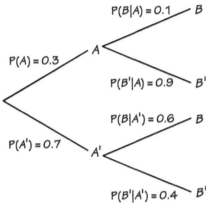

P(B|A) = 0.1 → B

P(A) = 0.3

A

P(B'|A) = 0.9 → B'

P(B|A') = 0.6 → B

P(A') = 0.7

A'

P(B'|A') = 0.4 → B'

(b) Probability of getting flu = $P(A) \times P(B|A) + P(A') \times P(B|A')$
$$= 0.3 \times 0.1 + 0.7 \times 0.6$$
$$= 0.45$$

(c) $P(A|B) = \dfrac{P(A \cap B)}{P(B)}$

$$= \dfrac{0.3 \times 0.1}{0.45}$$

$$= 0.067 \ (3 \text{ s.f.})$$

2 (a) Mean, $E(X) = \frac{1}{2}(a + b) = \frac{1}{2}(2 + 6) = 4$ minutes

(b) $Var(X) = \frac{1}{12}(b - a)^2 = \frac{1}{12}(6 - 2)^2 = 1.33$ minutes

(c) $P(c \le X \le d) = \dfrac{d - c}{b - a}$

Hence $P(3 \le X \le 4) = \dfrac{4 - 3}{6 - 2}$

$$= 0.25 \text{ minutes}$$

3 (a) Let X be the random variable 'weight of a sack of bird food'.
X is normally distributed with mean 12.2 kg and variance σ^2,
so $X \sim N(12.2, \sigma^2)$.

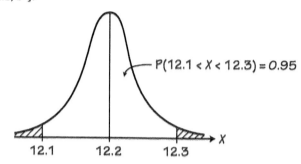

P(12.1 < X < 12.3) = 0.95

> Notice that the graph is symmetrical so the probabilities for $X > 12.3$ and $X < 12.1$ are the same.

From the above sketch the probability in each of the two tails $= \dfrac{(1-0.95)}{2}$

$$= 0.025$$

We can just look at the upper-tail to work out the z-values.
Drawing a quick sketch, of the standard normal distribution we obtain:

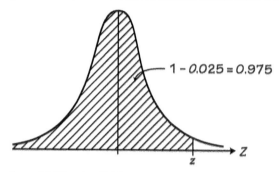

1 − 0.025 = 0.975

We now look up 0.975 using Table 4 Percentage Points of the Normal Distribution and find that the z-value is 1.960.
Now we need to find the value of σ for $X = 12.3$

> Note that there are two z-values (i.e. ±1.960) but we only need to use one of them.

$$z = \frac{x - \mu}{\sigma}$$

$$1.960 = \frac{12.3 - 12.2}{\sigma}$$

Hence $\sigma = 0.0510 \ldots$
So variance $\sigma^2 = 0.0026$ (2 s.f.)

(b) Sample mean $= \dfrac{183.8}{15} = 12.2533$

Null hypothesis, $\mathbf{H_0} : \mu = 12.2$
Alternative hypothesis, $\mathbf{H_1} : \mu \neq 12.2$

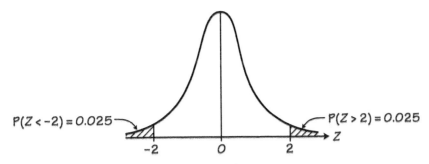

$$P(Z < 2) = 1 - 0.025 = 0.975$$

Looking up a probability of 0.975 in Table 4 Percentage Points of the Normal Distribution, we have $z = 1.960$

We also have $z = -1.960$

$$z = \frac{\bar{X} - \mu}{\frac{\sigma}{\sqrt{n}}}$$

$$1.960 = \frac{\bar{X} - 12.2}{\frac{0.05}{\sqrt{15}}}$$

$$\bar{X} = 12.2253$$

$$-1.960 = \frac{\bar{X} - 12.2}{\frac{0.05}{\sqrt{15}}}$$

$$\bar{X} = 12.1747$$

Hence, the critical region is $\bar{X} \le 12.1747$ or $\bar{X} \ge 12.2253$

The test statistic (i.e. the sample mean of 12.2533) lies inside the critical region, so there is sufficient evidence at the 5% level of significance that the mean weight of bird food has changed from 12.2 kg.

4 (a) (i) $r = 0.9$

(ii) $r = -0.6$

(iii) $r = 0$

(iv) $r = -1$

(b) $H_0 : \rho = 0$, $H_1 : \rho \ne 0$

This is a two-tailed test, so looking up 10% significance level for a two-tail test and $n = 40$, the critical value is read off from the 'Critical values of the product moment correlation coefficient' table.

From the table $\rho = \pm 0.2638$

Hence the critical value $= -0.2638$

Now the PMCC $= -0.55$ and as $-0.55 < -0.2638$ we can say the result is significant.

Hence there is evidence to suggest at the 10% level of significance that the two quantities X and Y are correlated.

Section B – Mechanics

1 Need to first draw a sketch of the graph.

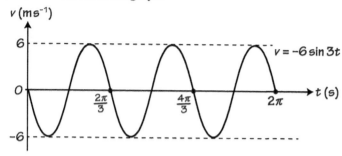

Notice that the graph is a reflection of $v = 6 \sin 3t$ in the x-axis and that the maximum and minimum values are $+6$ and -6 and notice that the normal sine graph is repeated 3 times in 2π owing to the $3t$ in the sine function.

To find the distance we can integrate the function between 0 and $\frac{\pi}{3}$ and then change the answer to a positive number and then double it.

Hence
$$s = \int_0^{\frac{\pi}{3}} \left(-6 \sin 3t\right) dt$$

$$= \left[\frac{6 \cos 3t}{3}\right]_0^{\frac{\pi}{3}}$$

$$= \left[2 \cos 3t\right]_0^{\frac{\pi}{3}}$$

$$= \left[(2 \cos \pi) - (2 \cos 0)\right]$$

$$= -2 - 2 = -4$$

We disregard the negative sign as we are only concerned with the distance.

The other section of the curve will have the same area and hence the same distance.

Hence total distance $= 2 \times 4 = 8$ m

> Note that this area is negative as this section of the curve is below the x-axis.

2 (a) $\mathbf{v} = 2 \sin 2t\mathbf{i} - 12 \cos 3t\mathbf{j}$

$$\mathbf{a} = \frac{d\mathbf{v}}{dt} = 4 \cos 2t\mathbf{i} + 36 \sin 3t\mathbf{j}$$

(b) $\mathbf{s} = \int \mathbf{v} \, dt = \int_0^{\frac{\pi}{2}} \left(2 \sin 2t\mathbf{i} - 12 \cos 3t\mathbf{j}\right) dt$

$$= \left[-\cos 2t\mathbf{i} - 4 \sin 3t\mathbf{j}\right]_0^{\frac{\pi}{2}}$$

$$= \left[\left(-\cos \pi\mathbf{i} - 4 \sin \frac{3\pi}{2}\mathbf{j}\right) - (-\cos 0\mathbf{i} - 4 \sin 0\mathbf{j})\right]$$

$$= \left[(\mathbf{i} + 4\mathbf{j}) - (-\mathbf{i} - 0\mathbf{j})\right]$$

$$= (2\mathbf{i} + 4\mathbf{j}) \text{ m}$$

Position vector $= (2\mathbf{i} + 3\mathbf{j}) + (2\mathbf{i} + 4\mathbf{j})$

$$= (4\mathbf{i} + 7\mathbf{j}) \text{ m}$$

> Note that this is the displacement vector. We now need to add the position vector of the starting point to this to find the position vector of the final position.

3 (a)

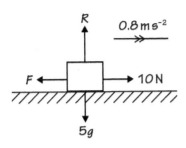

(b) Resolving vertically, we obtain $R = 5g$

$$= 49 \text{ N}$$

(c) Applying Newton's 2nd law, we obtain:

$$ma = 10 - F$$
$$5 \times 0.8 = 10 - F$$
$$F = 6 \text{ N}$$

(d) $F_{MAX} = \mu R$

$$6 = 49\mu$$
$$\mu = 0.1 \text{ (1 d.p.)}$$

> Note that as the box is moving it experiences the maximum frictional force F_{MAX}.

4 (a)

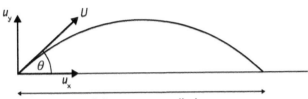

Horizontal distance travelled = range

To find the time of flight, we consider the vertical motion and equate the vertical displacement to zero. We will consider the upward direction as the positive direction for the motion.

$$s = ut + \frac{1}{2}at^2$$

$$s = U \sin \theta t - \frac{1}{2}gt^2$$

As $s = 0$, $0 = U \sin \theta t - \frac{1}{2}gt^2$

> Note that g is negative as it acts downwards.

Hence $t\left(U \sin \theta - \frac{1}{2}gt\right) = 0$, so either $t = 0$ or $U \sin \theta - \frac{1}{2}gt = 0$.

The $t = 0$, is when the particle is projected, so we use the other solution.

Hence, time of flight, $t = \dfrac{2U \sin \theta}{g}$

(b) Range = horizontal velocity × time of flight

Now horizontal velocity $= U \cos \theta$, and time of flight $= \dfrac{2U \sin \theta}{g}$ so substituting these into the above formula, we obtain:

$$\text{Range} = U \cos \theta \frac{2U \sin \theta}{g}$$

$$= \frac{2U^2 \sin \theta \cos \theta}{g}$$

Now $2 \sin \theta \cos \theta \equiv \sin 2\theta$

$$\text{Range} = \frac{U^2 \sin 2\theta}{g}$$

> Remember that this is the double angle formula.

(c) No frictional forces act (i.e. the only force acting is gravity).

(d) (i) Range $= \dfrac{U^2 \sin 2\theta}{g}$

$32 = \dfrac{U^2 \sin 120°}{9.8}$

$U = 19.0293$

$ = 19 \text{ m s}^{-1}$ (2 s.f.)

(ii) Time to reach maximum height $= \dfrac{1}{2} \times$ time of flight

$= \dfrac{1}{2} \times \dfrac{2U \sin \theta}{g}$

$= \dfrac{U \sin \theta}{g}$

$= \dfrac{19.0293 \sin 60°}{9.8}$

$= 1.6816 \text{ s}$

$= 1.7 \text{ s}$ (2 s.f.)

(iii) At the greatest height $v = 0 \text{ m s}^{-1}$, $u = U \sin 60$, $a = g = -9.8 \text{ m s}^{-2}$ and $s = ?$

$v^2 = u^2 + 2as$

$0^2 = (19.0293 \sin 60)^2 - 2 \times 9.8s$

$s = 13.8564 \text{ m}$

$= 14 \text{ m}$ (2 s.f.)

5 (a) $\dfrac{dP}{dt} \propto \sqrt{P}$

$\dfrac{dP}{dt} = k\sqrt{P}$

Separating variables and integrating, we obtain:

$\displaystyle\int \dfrac{dP}{\sqrt{P}} = k \int dt$

$\displaystyle\int P^{-\frac{1}{2}} dP = k \int dt$

$\dfrac{P^{\frac{1}{2}}}{\frac{1}{2}} = kt + c$

$2P^{\frac{1}{2}} = kt + c$

$2\sqrt{P} = kt + c$

When $t = 0$, $P = 1600$ so $2\sqrt{1600} = 0 + c$

Hence $c = 80$

Equation can be written as $2\sqrt{P} = kt + 80$

So $\sqrt{P} = \dfrac{kt + 80}{2}$

$P = \left(\dfrac{kt + 80}{2}\right)^2$

(b) When $t = 4$ and $k = 0.25$, $P = \left(\dfrac{0.25 \times 4 + 80}{2}\right)^2$

$= 1640.25$

$= 1640$ (nearest integer)